Copyright © 2016 by Onyeka I Uzoma MD

All rights reserved. This book or any portion thereof may not be reproduced or used in any manner whatsoever without the express written permission of the publisher except for the use of brief quotations in a book review.

Printed in the United States of America

First Printing, 2016

ISBN-13: 978-1535592154

ISBN-10: 153559215X

BISAC: Science / Philosophy & Social Aspects

Written and Designed by Onyeka Uzoma MD

Author contact: drbiggie2001@gmail.com

Introduction

Studying humans is most interesting. I really believe that the average anthropologist, sociologist and psychologist gets a real kick out of their work. That's just my opinion by the way. Have you ever been to a train station, coffee shop or any other public place for that matter and just allowed yourself the opportunity to be a passive observer of the dynamics of human interactions? I have done this quite a bit. So this in many ways set me on a journey to try to understand the very nature of man and our universe. The body of knowledge that is currently available is quite intimidating but I decided to try to make some kind of sense of it all.

I have examined various aspects of our common existence, from religion to anthropology to quantum mechanics to neurobiology to general relativity to a whole range of other branches of science.

What is my purpose? The truth of it has two sides; the first, to find a recurring pattern or a common theme within a plethora of diverse fields pointing us to the existence or non-existence of a higher intelligence in operation at all levels of the universe. What some may refer to as God or a Creator. Secondly, I know for a fact that learning to step out of one's intellectual comfort zone tends to change the way we process information, so in doing my research I broadened my worldview and tested my own sacred belief system, all in a bid to pass on a work of nonfiction that is enjoyable, readable and simple enough not to feel too academic.

So come with me, let's take a walk through four dimensional space-time while searching for physical/spiritual reality (whatever that may be).

CHAPTER ONE

Awakening: Wired to search for explanations.

If science is correct our universe is almost fourteen billion years old. Its size is almost impossible to fully comprehend, with over 100 billion galaxies, exotic supernovas, black holes, mysterious dark matter, enigmatic dark energy, a background of microwave radiation and other wonderful seen and unseen objects. Within this gargantuan realm lies a sparkling white colored galaxy called the Milky Way with no less than two hundred billion stars revolving round a central point. One of these stars- formed 4.6 billion years ago- is peripherally located. It is a yellow dwarf star (some dwarf, it weighs 2 with 30 zeros after it, in kilograms) and is called the Sun. It has 8 planets, dwarf planets and small solar system bodies revolving around it. These form our solar system and within a special area called its circumstellar habitable zone, or Goldilocks zone, lies our home planet; the Earth. The concept of a circumstellar habitable zone (CHZ) in astronomy, is a range of orbits around a star within which a planet's surface can support liquid water at sufficient atmospheric pressure. This region is also called the Goldilocks zone, a phrase derived from the fairy tale of *Goldilocks and the Three Bears*, in which a little girl chooses from sets of three items, ignoring the ones that are too extreme and settling for the one in the middle, which is "just right."

On such a grand scale one can't help but notice how infinitesimally minuscule our place in the universe actually is.

Now let's get slightly less numeric and a little more organic as we proceed.

The dawn of humankind, as we know it, began with the emergence of a species called *Homo sapiens* if we are to go strictly by what scientists have documented. The predecessors of modern humans were far less intelligent and so they eventually became extinct. Tough world I guess or maybe just a true reflection of Darwin's survival of the fittest concept of evolution. By the way, many people don't believe in evolution, I thought I should just mention that right at this point. Current scientific research has two leading hypotheses concerning the origin of modern humans; the recent African origin and the multiregional origin hypotheses. The former is the mainstream model and it postulates that the humans we see today evolved solely on the African continent (from now extinct primates) between 100,000 – 200,000 years ago, then migrated out of Africa during two different waves of migration. The hypothesis of multiregional origin suggests the evolution of modern man was a worldwide event.

So whatever happened to Adam and Eve?

If we look at the account of creation and the appearance of man as cataloged in the Biblical book of Genesis, it describes the creation of the first humans who were named Adam and Eve. Is there any scientific evidence that points to the possibility of modern mankind having a common male and female starting point or ancestry if you like? Surprising as it may sound, the answer is yes.

A lot concerning human history can be deciphered using parts of the human genome. Two parts to be more specific: the Y chromosome and mitochondrial DNA. Evolution affects most cellular material, even certain DNA sequences, and causes them to change over time. Fortunately, the aforementioned parts (the Y chromosome and mitochondrial DNA) are unaffected; they are passed down unchanged or intact. According to current hypothesis every human living today inherited the same mitochondrial DNA from a woman referred to as Mitochondrial Eve, who lived in Africa possibly 160,000 years ago. Another interesting piece of information appears to be found in the fact that all men currently living on the planet today seem to have inherited their Y chromosome from a single male ancestor who lived between 200,000 – 300,000 years ago most probably in Africa. He has been aptly named Y – chromosomal Adam or the Y-chromosomal most recent common ancestor.

No matter what you as an individual believe or do not believe, there are a few common denominators that cut across every school of thought and belief system. For now I will only zero in on two constants. The first is what I call the "Search for the Source" (SFS), there is something hard wired into the brain of man that has driven man even in his most primitive state to search for the source of his origin. Put differently, man has always been searching for God. At the very core of the human race, man has an intrinsic default state to look for God.

The second common denominator is what I will term the "Quest to Know" (QTK), this is the yearning that has existed in mankind to answer questions concerning its purpose, ask questions about why things are a certain way in the universe, the afterlife conundrum, etc.

The SFS denominator.

A cursory look at various time frames in mankind's history shows that any cluster of people who have or had a language always had or have a creation mythology (be it right, wrong or even outlandishly absurd) that usually has a God or multiple gods in it. Some ancient peoples worshiped the Sun, others the moon, trees, lightning, sacred animals, ancestors, spirits and even rocks. Why the search for a Creator in the first place? Is it vaguely possible that the Creator (if one exists) wired our brains to seek him out? This is just me thinking out loud.

However, I do find this most intriguing. Even the great scientific minds have used mathematics, physics and theories to search for God. Some admit to this, some are coy about the subject while others set out to validate the exact opposite. The views of many of the greatest scientific minds on the issue of the existence or non-existence of a Creator would certainly make for fascinating discourse. Broadly speaking their views are so varied that an entire book might not be able to pin down the essence of their varied perspectives on the subject matter, hence I will try to make only a few references to such in the course of this book as and when the need arises.

QTK

There is no doubt that mankind has from its infancy searched for answers to explain the events and things around it. This quest continues till this very day, along the way we developed religion and science as tools to help figure out these apparent mysteries. Science has indeed served us well but so has religion whenever it was applied constructively. We know quite a bit about ourselves and our environment thanks to advances in science and technology. However the fact that science itself is flawed is serious food for thought.

Science is flawed? Did some gasp at the notion of science being anything but perfect? Yes, it is and as we go along the facts should speak for themselves.

Religion on the other hand is not without its own pitfalls, all the same it has helped us create societal as well as individual moral compasses, it has also served as a template from which many of our modern laws were derived. In this way it provided structure in the face of a potentially chaotic and violent society. Without order in the society science might most likely have never been able to thrive, even though many a times there has existed conflict between both science and religion. As paradoxical as it might appear I believe they are reverse sides of a two faced coin.

The Milky Way Galaxy

Reference

1. Hall, Shannon (2015-05-04). "Size of the Milky Way Upgraded, Solving Galaxy Puzzle". Space.com. Retrieved 2016-08-08.

2. NASA. The Goldilocks zone (10-02-2003). NASA science/news. Science.nasa.gov. Retrieved 07-24-2016.

3, Kopparapu, Ravi Kumar (2013).. "A revised estimate of the occurrence rate of terrestrial planets in the habitable zones around kepler m-dwarfs". *The Astrophysical Journal Letters*. **767** (1): 8.

4, Major Events in the History of Life. Edited by J. William Schopf. Jones & Bartlett Publishers; 1st edition 1991 Pg 168.

5, Matthew H. Nitecki, Doris V. Nitecki. Origins of Anatomically Modern Humans. Springer, 31 January 1994.

6, **Stringer, Chris B.** (1994). [First published 1992]. "Evolution of Early Humans". In **Jones, Steve**; Martin, Robert D.; **Pilbeam, David**. *The Cambridge Encyclopedia of Human Evolution*. Foreword by Richard Dawkins (1st paperback ed.). Cambridge, UK: Cambridge University Press.

7. Wolpoff, MH; **Hawks, J**; Caspari, R (2000). "Multiregional, not multiple origins" (pdf). *American Journal of Physical Anthropology*. **112** (1): 129–36.

8. Blaine Bettinger (20 July 2007). "Mitochondrial Eve and Y-chromosomal Adam". *The Genetic Genealogist*.

9. Pedro Soares et al 2009, **Correcting for Purifying Selection: An Improved Human Mitochondrial Molecular Clock** and its **Supplemental Data**. The American Journal of Human Genetics, Volume 84, Issue 6, 740–759, 4 June 2009.

Chapter Two

Origins: The roots of religion.

The truth is that it is difficult to find a strong dividing line between the origins of religion and that of science. Both have a common origin, they both were birthed from the unified womb of SFS and QTK. As highlighted in the opening chapter, mankind as it evolved needed answers to questions that could not readily be explained.

We will go off on a slight tangent using scientific data and theoretical modules to trace the origin of religion; religion being any set of practices, beliefs, rituals and rules used to worship a God or gods. I have left out certain other facets of religion like the concept of "spirituality" on purpose.

Why? Because at this point I truly wish to avoid any form of ambiguity hence the need to focus on the simpler, but core aspects of the current subject matter.

Now back to the subject matter, to understand where religion came from we must go back in time, for the simple reason that religious practices emerged thousands of years ago in the distant past of mankind.

The evolution of man over time separated him from his closest living relatives, the chimpanzee and the bonobo. The split from a common ancestor is believed to have occurred six to eight million years ago.

This evolutionary split allowed humans to develop along a slightly different trajectory. The most important change was that of increased brain size, there was an approximate tripling in human brain size from the time of the evolutionary fork in the road till about half a million years ago. This is evident when examining an early ancestor of modern man- *Australopithecus afarensis*- the species to which the iconic early human fossil "Lucy" belonged to. This ancient ancestor of modern humans had a brain that was less than 500 cubic centimeters in volume, or about a third the size of the modern human brain. **The greater part of the increased brain size involved an area called the neocortex. This is the newest evolutionary part of the brain and forms the top layer of the cerebral hemispheres. Thanks to the larger neocortex, mankind became equipped with the tools for complex thought patterns, conscious thought (by extension self awareness), and language.** Surprisingly this increase might have been sparked by a single gene unique to humans according to a study led by geneticist Marta Florio, at the Max Planck Institute of Molecular Cell Biology and Genetics in Dresden. The insertion of this gene in rodents produced marked enlargement in brain size, lending credence to this genes possible role in triggering brain enlargement in humans.

The very traits that were needed to permit the emergence of religion like: realization of self, high intelligence, an ability to co-ordinate symbolic communication, a concept of social norms plus a notion of continuity, were all **now within the expanding repertoire of human brain power. Some experts suggest that religion** may have sprung up- during evolutionary twists and turns- as a means of cementing harmony within groups of humans. Who having figured out the inevitability of mortality on a personal scale looked for a means of having a shared belief system that offered the hope of continuity of self in some shape or form.

As humans began to use the most primitive of tools- before progressing to develop even more complex tools- it is believed that as this happened humans began to grasp the concept of causality. This appears logical, for one must comprehend how a tool is used before putting it to use which by extension requires at least a basic understanding of cause and effect.

With greater brain size and brain power came flashes of insight.

Religion requires a pattern of symbolic interaction/communication, such as language, in order to be transmitted from one individual to another.

With the development of language, humans could articulate their thoughts, fears, reflections and literally discuss them with others.

Based on the above premise science writer Nicholas Wade stated: "Like most behaviors that are found in societies throughout the world, religion must have been present in the ancestral human population before the dispersal from Africa 50,000 years ago. Although religious rituals usually involve dance and music, they are also very verbal, since the sacred truths have to be stated. If so, religion, at least in its modern form, cannot predate the emergence of language. It has been argued earlier that language attained its modern state shortly before the exodus from Africa. If religion had to await the evolution of modern, articulate language, then it too would have emerged shortly before 50,000 years ago."

When we look at a behavioral pattern seen in our supposed distant relative the Chimpanzee, they cohabit in **groups** of about 50 members. It is a plausible notion that our early ancestors lived in groups of a similar size.

Over the course of time, evolutionary trends set up ever increasing community sizes. A greater need to achieve group cohesion would be an essential prerequisite. This could have been provided by a common belief system that established a code of conduct.

Within groups of greater than 100 individuals, morality might have evolved as a tool for social control, conflict resolution and enhanced group bonding. Some psychologists contend that religion emerged after morality and strengthened it by bringing in supernatural agents as a means of increasing the social scrutiny of individual behavior.
By including ever-watchful ancestors, spirits and gods in the social realm, humans discovered a means that was effective in taming the extremes of self centered behaviors while building communal affiliation. This adaptive attribute of religion would have impacted positively on collective survival.

The current consensus within the ranks of cognitive scientists is that religion is an offshoot of improved brain architecture as it evolved. However, dissenting voices appear when it comes to the precise mechanisms that orchestrated the evolution of the religious mind.
Two different schools of thought propose that either religion evolved as a consequence of natural selection and as such has a selective advantage, or that religion is an evolutionary byproduct of other mental adaptations.
To analyze these two differing views in simpler terms, the first group (school of thought) believes that in the harsh reality of "only the strong survive" that those humans who had a religious viewpoint or outlook were more likely to survive because they would most likely work

together, increasing their chances of survival and further evolutionary development. The second group are of the opinion that religion just happened to happen as the mind adjusted over time to the new realities that came its way. It might have been either of the two or even a combination of both that led to the above scenario.

The mechanisms that propelled the mind toward religion may have included; i) the ability to create causal narratives for natural events, ii) the capacity to pinpoint the presence of things that might cause harm, and iii) the ability to articulate the fact that other individuals have minds of their own with their own beliefs, desires and intentions. Three of these adaptive features have enabled humans to think up agents responsible for many observations that could not readily be explained otherwise, e.g. the cycle of the moon, lightning, thunder, stars, the complexity of life, etc. The appearance of collective religious belief identified the agents as deities that made these explanations believable.

Some scholars like geneticist Dean Hamer have suggested that religion is genetically "hardwired" into the human condition, one example being; the God gene hypothesis, which states that some variants of a specific gene predispose humans to spirituality. The gene has even been identified as the **vesicular monoamine transporter 2 (VMAT2)** gene. Hamer opined that the presence of a God gene or God genes is a hallmark of the Creator's ingenuity.
However critics like **Carl Zimmer** disagree with this notion of VMAT2 being the God gene. Zimmer claimed that VMAT2 can be described as a gene that is responsible for less than one percent of the variance of self-transcendence scores. Self-transcendence being a personality trait associated with experiencing spiritual idea like considering oneself an integral part of the universe.

Okay, could there be genes that make you want to get on your knees and worship? As far-fetched as it may sound there is a possibility that it might just be very sound science. If indeed the possible genetic underpinnings of religion are researched to a point of certainty. This would point to the validity of the notion that religion is an intrinsic attribute of humanity, whose expression would certainly be affected by social as well as environmental factors.

Another view is that religion amongst humans arose from an increase in quantity and activity of a chemical called dopamine in the human brain. Dopamine promotes an appreciation of distant space and time, which is essential for the firm entrenchment of religious experience.

Dopamine might play a role in having a religious appreciation of events but to say or think that a single chemical in the brain is responsible for the emergence of religion is certainly extremely unlikely and doesn't rest on any sound scientific evidence.

Now where did all those religious symbols come from anyway?

The use of symbolism in religion features prominently as it helps represent supernatural beings and ideas. Symbolism most probably appeared alongside the ability of humans to communicate through drawings and other forms of art.

Some of the earliest evidence of symbolic behavior is associated with **Middle Stone Age** sites in Africa. From at least 100,000 years ago, there is evidence of the use of pigments such as **red ochre. It was used** in rock art paintings and also at burial sites.

The significance of this shows up in the argument that the color red is universally seen in human religion as being a representation of blood, sex, life and death.

The appearance of religious texts marked the beginning of religious history, a direct byproduct of the new human skill called writing.

The *Pyramid Texts* from ancient Egypt are amongst the oldest known religious texts in the world, dating to between 2400–2300 BCE. Writing played a pivotal role in sustaining and spreading organized religion. In groups and societies that were pre-literate, religious ideas were based on an oral tradition, the contents of which were articulated by high priests or shamans and were dependent on the collective memory of the practitioners. With the advent of writing, information that was not easy to remember could easily be stored in sacred texts that were maintained by a select group. Writing enabled religions to develop coherent doctrinal systems.

Multiple gods or one God?

Polytheism, the belief in multiple gods, was the typical form of religion during the **Bronze Age** and Iron Age. It is most probable that polytheism evolved into monotheism- belief in one god- in those religions where only one god is recognized. This hypothesis however may not apply to all religions that have a single god.

Some religions have elements of both the concept of a single god and that of multiple gods. Hinduism for example, has elements of the two. In fact some scholars view the different practices within Hinduism as separate religions that have been wrongly placed under one big umbrella.

The belief in one god instead of a plethora of gods is a feature seen in various religions; the Abrahamic religions (Judaism, Christianity and Islam), Sikhism, Deism, **Babism**, the Bahà'ì faith, Eckankar, Tenrikyo, Shaivism and Mandaeism.

Elements of the worship of a single God can also be found in ancient Chinese religion as well as indigenous African religion. For instance the faith system held by most dynasties of China since at least 1766 BCE until the modern period focused on the worship of "shangdi" (literally "Above Sovereign", also translated as "God") or Heaven as an omnipotent force. However it did recognize the existence other lesser gods.

In Africa, a supreme God called Chukwu was worshiped among the Ibo as their supreme God; other lesser deities were also recognized.

Tracing the origin of religion to death rituals has also been debated and considered as a possible beginning for religious practices. However pointing to death rituals like burials as a possible starting point for religion has its flaws as death rituals are not seen exclusively in humans, they are seen among certain other animals. Across cultures humans find some form of value in guarding or watching over the body of the deceased. Though the details vary from tradition to tradition, the pattern is certainly undeniable.

These behaviors are by no means unique to humans, various elements of these behaviors have been observed and documented in elephants, dolphins, and chimpanzees. They too have been observed to exhibit attachment to their dead and tend to guard the carcasses of their departed.

The simple argument that works strongly against religion having its roots in death rituals would be the straightforward question- how come these other animals never evolved beyond rudimentary death rituals, why didn't they go further to develop more complex religious practices? The answer in my opinion would be the possible presence of an instinctive trait that exists in all these animals including humans to "respect" the dead and grieve the loss of one of their kind.

In exploring religion and its roots, the need exists to give a brief overview of the world's largest religious group, Christianity. Why? Because curiosity draws us to certain preconceived notions that may or may not be correct. A simple example would be the notion that the largest religion on the planet might have features that are absolutely unique to it or that there is the likelihood that the correct religion must be situated where the majority of humans have pitched their religious tents. In both cases such notions are based on the mind playing the game of numbers as opposed to deep analytical thinking that is devoid of bias.

Christianity is the world's largest religion with well over 2 billion adherents.

Here the concept of Trinity is espoused; three beings forming the Godhead (God the Father, God the Son and God the Holy Spirit). It is an offshoot of Judaism. The Bible serves as its holy book and is believed to be the inspired word of God. The Bible has two parts, the New Testament and the Old Testament.

The central theme in Christianity is that God the Father sent His only begotten son Jesus to the world to die in order to save a fallen world and ultimately redeem mankind. According to the bible the first humans- Adam and Eve had violated a most sacred commandment, this violation brought about a separation between God and humanity. Jesus restored this relationship by dying, then resurrecting and ascending into Heaven, belief in Jesus and following his path as well as his teachings guarantees a place in paradise in the afterlife as a reward. On the other hand rejection of him and living contrary to the values he espoused results in eternal damnation when you die.

The one strong similarity here between the biblical account of human history and that of science is the presence of a common male and female ancestor for all living humans. A second similarity is that they both clearly spell out a starting point for the beginning of time/physical reality and predict an end of time. Other than these the two tend to be quite divergent.

Now that we have taken an in-depth look at the likely origins of religion, what questions come to mind?

So many for me but here are just a few of them; is religion an evolutionary response to humanity's ever changing realities? Here it would be plausible to say yes but that would only be a possibility if you believe in evolution.

Why should anyone buy into the scientific view of the emergence of religion, when most religions outline their own histories in the sacred texts that are peculiar to them? Now it is difficult for me to take an unequivocal stand on this as I am both a man of science and one of faith. So it's up to you to evaluate the question and reach a conclusion based on your own world view.

Irrespective of location, humanity embraced different forms of a common behavioral pattern- religion, the big question is why? I would put it down to SFS and QTK, pure and simple.

Are we wired to search out a Creator (if one exists) through religion as a medium of communication between us and the Creator? I would say a categorical yes to this question as religion is seen universally on a global scale cutting across every known culture.

Is religion a flawed aspect of our evolution or does it tell us more about who we really are? In many ways it tells us a great deal about who we are and how we have come to this point in our existence, be it evolutionary or in the laws we have drawn from religion to govern ourselves. The sense of what is right versus what is wrong are mirrors of our religious leanings. For example, honor killings of family members that are carried out on the Indian subcontinent might be viewed by the executors as being a noble act and a protection of a family's dignity, where as this would qualify as premeditated murder in America and many parts of western Europe. The two divergent views can be traced to both religious and cultural beliefs. Cultural practices we must remember tend to be heavily influenced by religion.

Why has religion continued to thrive despite new scientific explanations for many natural events that previously seemed to apparently have only a religious basis?

The questions go on and on, a few have been deliberately left open ended, however I believe at some point in this our journey we might be blessed (yes, I did indeed say blessed) with greater insight and possibly answers will emerge.

Reference

1. King, Barbara (2007). Evolving God: A Provocative View on the Origins of Religion. Doubleday Publishing." ISBN 0-385-52155-3.
2. Tyson, Peter (July 1, 2008). "Meet Your Ancestors". *NOVA scienceNOW*. PBS; WGBH Educational Foundation. Retrieved 2015-04-18.
3. Ehrlich, Paul (2000). *Human Natures: Genes, Cultures, and the Human Prospect*. Washington, D.C.: Island Press. p. 214. ISBN 1-55963-779-X.
4. Dávid-Barrett, T.; Dunbar, R. I. M. (2013-08-22). "Processing power limits social group size: computational evidence for the cognitive costs of sociality". *Proceedings of the Royal Society of London B: Biological Sciences*. **280** (1765): 20131151.doi:10.1098/rspb.2013.1151.
5. Marta Florio. Human-specific gene *ARHGAP11B* promotes basal progenitor amplification and neocortex expansion. Science 27 Mar 2015:Vol. 347, Issue 6229, pp. 1465-1470 DOI: 10.1126/science.aaa1975.
6. Stephen Jay Gould, Paul McGarr, Steven Peter Russell Rose (2007). "Challenges to Neo-Darwinism and Their Meaning for a Revised View of Human Consciousness". *The richness of life: the essential Stephen Jay Gould*. W. W. Norton & Company. pp. 232–233.
7. Lieberman (1991). *Uniquely Human*. Cambridge, Mass.: Harvard University Press. ISBN 0-674-92183-6.
8. Matthew Rutherford. The Evolution of Morality. University of Glasgow. 2007. Retrieved June 6, 2008.
9. Rossano, Matt (2007). "Supernaturalizing Social Life: Religion and the Evolution of Human Cooperation.
10. Atran. S, Norenzayan. A (2004). "Religion's evolutionary landscape: counterintuition, commitment, compassion, communion.". *Behavioral and Brain Sciences*. **27** (6): 713–30.
11. Dávid-Barrett, Tamás, Carney, James (2015-08-14). "The deification of historical figures and the emergence of priesthoods as a solution to a network coordination problem". *Religion, Brain & Behavior*. **0** (0): 1–11.
12. Steven Pinker 2005. The evolutionary psychology of religion. Freehold today vol22 no.1.
13. Lionel Tiger and Michael McGuire (2010). *God's Brain*. Prometheus Books. ISBN 978-1-61614-164-6.

14. Previc, F.H. (2009). The dopaminergic mind in human evolution and history. New York: Cambridge University Press.
15. Robin Ian MacDonald Dunbar, Chris Knight. The Evolution of Culture: An Interdisciplinary View. Edinburgh University Press, 1999.
16. Andre Leroi-Gourhan, Annette Michelson. *The Religion of the Caves: Magic or Metaphysics?*, 1986 October 37, The MIT Press, pp. 6-17.
17. Budge, Wallis (1997). *An Introduction to Ancient Egyptian Literature*. Mineola, N.Y.: Dover Publications. p. 9. ISBN 0-486-29502-8.
18. Cross, F.L.; Livingstone, E.A., eds. (1974). "Monotheism". The Oxford Dictionary of the Christian Church (2 ed.). Oxford: Oxford University Press.
19. Frazier, Jessica (2011). The Continuum companion to Hindu studies. London: Continuum. pp. 1–15. ISBN 978-0-8264-9966-0.
20. Monotheism", *Britannica*, 15th ed. (1986), **8**:266.
21. Zhao, Yanxia. *Chinese Religion: A Contextual Approach.* 2010.
22. Egboh, Edmund O. (1972). "A Reassessment of the Concept of Ibo Traditional Religion". *Numen*. **19** (1): 68.doi:10.2307/3269588.
23. O'Connell, Caitlin (2007). The Elephant's Secret Sense: The Hidden Lives of the Wild Herds of Africa. New York City: Simon & Schuster. pp. 174, 184.
24. van Leeuwen, E. J.C., Mulenga, I. C., Bodamer, M. D. and Cronin, K. A. (2016), Chimpanzees' responses to the dead body of a 9-year-old group member. Am. J. Primatol., 78: 914–922. doi:10.1002/ajp.22560.
25. ANALYSIS (19 December 2011). "Global Christianity". Pew Research Center. Retrieved 17 August 2012.

Chapter Three

Understanding Science: its beginnings, beauty and imperfections.

The evolution of Science

The mention of the term science evokes different images in the minds of people, but what really is science?
The word science comes from the Latin expression "scientia," meaning knowledge.
Two different definitions of this term that capture its entire essence are outlined thus;
1) Science is a systematic enterprise that builds and
organizes **knowledge** in the form of
testable explanations and predictions about the universe.
2) Science is a body of **empirical**, theoretical, and **practical** knowledge about the natural **world**, produced by individuals who emphasize the observation, **explanation**, and prediction of real world phenomena.
These are to a large extent self explanatory statements without any need for further elucidation, the two complement one another without being contradictory.

The roots of science go thousands of years back, with seemingly rudimentary ideas undergoing a process of scrutiny, reevaluation and subsequent building on.

In the earliest years of humanity, knowledge of any form was passed on from one generation to the next through oral tradition. An example being *Forest gardening*, a plant-based food production system that may be the world's oldest agro-ecosystem. It began in prehistoric times in the wet foothills of monsoon regions and along jungle-clad river banks. In the gradual process of a family improving their immediate environment, useful tree and vine species were identified, protected and improved whilst undesirable species were eliminated. Evidence of such plant cultivation dating from 23,000 years ago has been found at the Ohalo II site.

This was prior to the appearance of writing, the writing of language is thought to have appeared independently in two different places; Mesopotamia around 3,200 BC and Mesoamerica around 600 BC. So this early form of agriculture could only have been sustained through oral teaching as well as observation.

The fact that astronomy was studied in the pre-literate era as shown by archeological evidence further shows that a pattern of oral transfer of knowledge must have existed prior to the advent of writing.

With the emergence of writing came the ability for knowledge to be better transmitted across generations with superior accuracy and reproducibility.

The advancement of agriculture provided better supply of food. This in turn freed up time for other tasks. One such advancement in farming seen in ancient times was the development of irrigation in Mesopotamia and ancient Egypt.

With this enhanced food security, the traditional communal hunter-gatherer social structure could now focus its intellectual energies differently; as the existence of humanity had registered a shift from primarily focusing on self preservation through the search for nourishment to a society where food became more readily available. These intellectual energies were partly shifted to include systematic studies of nature and the study and appraisal of written information recorded by others.

A timeline on the progress and growth of science can be seen by studying the areas where and individuals who set up the initial building blocks of scientific thought and methods.

Science in Africa, Mesopotamia and Greece.

The development of geometry occurred in ancient Egypt, this is in addition to major advances recorded in mathematics and astronomy. In medicine, techniques adopted by those who administered healthcare practices in Egypt are still seen today namely; examination, diagnosis, treatment, and prognosis.

The solar year and lunar month have their origins in the work of early Mesopotamian astronomers.

The Greek philosophers of yore made inroads that are considered by some as being the pivotal crucible from which the advancement of early science was made possible. Plato and Aristotle developed *deductive reasoning* which was found to be extremely useful in future scientific inquiry. Empiricism and the idea that the use of observation and induction could be used to arrive at universal truths was birthed in Greece by Aristotle. His writings would go on to influence the scholarly activities of later centuries.

Other Scientific hot beds

India, China, Islamic scholars and Europe (in the middle ages) all made significant contributions to the body of early scientific knowledge.

Aryabhata, an Indian astronomer and mathematician in his writing: Aryabhatiya (in the fifth century) introduced trigonometric functions like sine, cosine and inverse sine. It also included algorithms of algebra. Less than two centuries later his compatriot, Brahmagupta suggested that gravity was a force of attraction. Many years later, Sir Isaac Newton would greatly expand our knowledge of gravity in his work *Philosophiae Naturalis Principia Mathematica* released in 1687. Newton's work explained the orbit of planets round the Sun.

The ancient Chinese made notable strides in mathematics, with the use of negative numbers and decimal fractions by the first century BCE. In seismology, Zhang Heng invented a seismometer in 132 CE which served as an early warning system for the authorities in the capital Luoyang whenever an earthquake had occurred in a location indicated by a specific cardinal or ordinal direction pointed out by Zhang's novel invention.

The use of experiments to distinguish between competing theories was started by Islamic scientists.

With his work in optics, Ibn al-Haytham is considered the father of optics. More so with his empirical proof of the intromission theory of light.

The propagation of scientific knowledge in Europe was initially concentrated in monasteries. However by the twelfth century, Europe experienced the birth of medieval universities, these universities played a crucial role in furthering scientific study.

The Scientific Revolution
Most historians give the timeline for the Scientific Revolution as having begun around 1543 when works like *De Revolutionibus* by the astronomer Nicolaus Copernicus were firstprinted.

The period was considered a revolution as it was marked by a willingness to question previously held truths and a quest for fresh answers. It ultimately brought forth a period of major scientific advancement.

Copernicus in 1514 posited that the earth moved round the Sun. Nonetheless, this was not the first time such a model of heliocentrism had been put forward. Historically, the first record of this was documented by Aristarchus of Samos (310 – 230 BCE), a Greek astronomer and mathematician. Unfortunately not much attention was paid to the initial idea that the sun and not the earth was the center of the solar system. This could have in the earlier years been due to the fact that the majority of astronomers embraced the idea of the Earth being the center of the universe, a position that was endorsed many years later by the church.

The work of Copernicus was closely followed by that of Johannes Kepler, which made better the work of the former by pointing to an elliptical orbit of the planets rather than a circular one.

Italian astronomer Galileo Galilei endorsed Copernicus' model of the solar system.

The Catholic Church was very influential during the period Copernicus and Galileo published their works on heliocentrism, their books were later banned by the Church and labeled heretic. Copernicus died shortly after the release of his published work and as such faced no personal persecution from the Church. In 1610, Galileo published his work *Sidereus Nuncius (Starry Messenger)*, describing the observations he

had made with the new **telescope**, namely the phases of **Venus** and the **Galilean moons** of Jupiter. With these observations he promoted the **heliocentric** theory of **Copernicus**. Galileo's initial discoveries were met with stiff opposition within the Catholic Church, and in 1616 the Inquisition declared heliocentrism to be formally heretical.
The Inquisition was a group of institutions within the judicial system of the **Catholic Church** that aimed to combat **heresy**. Galileo was censored being ordered to refrain from holding, teaching or defending heliocentric ideas.
Sir Isaac Newton's work explained how the sun's gravity was responsible for the elliptical orbit of planets. The greatest success of Newton's theory was in the prediction of the existence Neptune based on the motions of Uranus. Newton invented calculus as well, it must also be acknowledged that calculus was also independently invented by **Gottfried Leibniz**. It later became an essential tool in much of the later development in most branches of physics.

As with some of his predecessors, Newton had his own critics. He was criticized for introducing occult agencies into science because of the postulate of an invisible force acting over vast distances.

It's interesting to note that any theory that shifted the paradigm away from a Creator centered view was met with different levels of opposition even from fellow scientists as recently as the time of Isaac Newton.
This was in many ways a natural response seen even today; people tend to resist change especially when it challenges their long held belief systems. I prefer to be open minded, searching instead for an intersection between new knowledge and the belief in a Creator.
Those who censored Galileo based on their perceived views of how biblical literature should be interpreted acted in ignorance as the same literature could be interpreted in diverse ways that would have easily

accommodated the evidence based findings of Galileo and by extension the work of Copernicus. This they failed to do in my opinion for reasons that bordered on maintaining the dominance and societal influence of the Catholic Church at the time.

Further scientific advances: a brief overview.

The breakthroughs in science as well as its advancement over time would take more than a single book to cover; it would be uncharitable to overlook the great contributions that appeared in various fields of science between the eighteenth and early twentieth century. However a brief mention of certain personalities and their contributions will highlight the expansion of the body of scientific knowledge over this time frame.

The Swiss mathematician **Daniel Bernoulli** (1700–1782) made important mathematical studies of the behavior of gases and has been referred to as the first mathematical physicist.

In 1776, **John Smeaton** published a paper on experiments relating power, **work, momentum, kinetic energy**, and supporting the **conservation of energy.**

In 1821, **Michael Faraday** built an electricity-powered motor, while **Georg Ohm** stated his law of electrical resistance in 1826, expressing the relationship between voltage, current, and resistance in an electric circuit.

In 1859, **James Clerk Maxwell** discovered the **distribution law of molecular velocities**. Maxwell showed that electric and magnetic fields are propagated outward from their source at a speed equal to that of light and that light is one of several kinds of electromagnetic radiation, differing only in frequency and wavelength from the others.

In the 19th century, experimenters began to detect unexpected forms of radiation: **Wilhelm Röntgen** caused a sensation with his discovery of X-rays in 1895; in 1896 **Henri Becquerel** discovered that certain kinds of matter emit radiation on their own accord.

Albert Einstein in 1905 released his theory of *special relativity*. It highlighted the implications for bodies moving at speeds close to that of light. Then in 1915, Einstein's theory of *general relativity* completely shifted the paradigm. This theory accurately predicted events that Newton's laws could not. For instance the accurate prediction of the orbit of mercury, which Newton's laws fell short of doing validated Einstein's general relativity. It also predicted the existence of gravitational waves.

A prediction that was confirmed a whole century later by the actual detection of gravitational waves in September 2015 by the twin Laser Interferometer Gravitational-wave Observatory (LIGO) detectors, located in Livingston, Louisiana, and Hanford, Washington. This marked the first time scientists had ever observed gravitational waves. The waves were arriving Earth from the final fraction of a second of the merger of two black holes to produce a single, more massive spinning black hole.

The significance of the discovery of gravitational waves is in the fact that it provides further evidence that strengthens the ingenuity of the theory of general relativity.

Most people are acquainted with Einstein's legendary formula; $E = mc^2$, where E is energy, m is mass and c is the speed of light in a vacuum. This apparently simple formula has far reaching implications. The conversion of small amounts of matter into energy is vividly demonstrated in the detonation of an atomic bomb. This equation in conjunction with a number of advanced scientific equations (like scattering cross sections and nuclear potentials) reveal that about an ounce of matter was all that may have been needed to produce the amount of energy needed for the destruction that followed the explosion of the bomb dropped on Hiroshima.

It takes an awesome amount of analytical reasoning to even come up with some of the mathematical calculations necessary to put forward a sound theory that can predict the existence of things that have neither been seen nor have any suggestive allusions as to their existence. The predictions of the *Standard Model* of particle physics, a theory that encompasses all of nature's subatomic particles, forces and interactions; it was a theory of the 1970s and it predicted the existence of a particle- the Higgs boson, occasionally referred to as the 'God particle' which is responsible for giving elementary particles their mass. The announcement of its discovery in 2012 was an important step in the consolidation of scientific understanding.

Although I think the reference to it as the 'God particle' hinges both on hyperbole and a subtle albeit inadvertent insight into the mindset of the proponents of the existence of the particle- the possible belief in the reality of a God.

The beauty of Science

Science has given us the power to influence our natural environment. It has made it possible through its closest sibling— technology for humanity to totally transform its existence, from a simple calculator to supercomputers, from the marvels of modern medicine to cutting edge robotics. It has simplified our lives, reduced childhood mortality from diseases that are prevented through vaccination, it made mechanized agriculture a means of sustaining the explosion in global population, it has effectively built invisible *bridges* and digital superhighways that have turned our world into what many have aptly called a global village.

Life on earth has changed from the days of the hunter–gather, who had to contend with the harsh vagaries of nature and the wild. His was a contentious existence compared to that of the average human living in present times. Though today's living conditions for many are still far from utopian, it is a gigantic leap forward from what once was existed.

The beautiful imagery of the universe produced by devices like the celestial based Hubble telescope serve as a pseudo–egotistical reminder of how far we have come. The James Webb space telescope which is billed to replace Hubble has immense potential. If it functions as planned we are in for some very profound insights into the beginnings of our universe. One particular goal involves observing some of the most distant objects in the **universe** currently beyond the reach of existing ground and space-based instruments, such as the **formation of the first galaxies**. Another goal is to understand the **formation of stars** and planets; this will include direct imaging of exoplanets.

This awe inspiring beauty of science has an ugly flip side, that includes an inexhaustible catalog of things most of us would wish either never happened or simply do not exist. The emergence of new drug resistant disease strains, German doctors who froze Russian prisoners to death to discover the effects of hypothermia on the human body, global warming, the proliferation of nuclear arms, cybercrime, the disappearing concept of personal privacy, chemical warfare, the use of human subjects to study the evolution of syphilis as a disease, the gas chambers of the Holocaust, and so many others. The very science that seeks to give us explanations may be the very thing that ends our existence.

It would be escapist to even think that some of the above examples were the product of monstrous acts in the name of science while others were crimes resulting from research with a clear aim, target or objective .

For instance let's examine a potential candidate for the latter- the *Tuskegee Study of Untreated Syphilis in the Negro Male*, otherwise known as the *Tuskegee Syphilis Experiment*, an infamous **clinical study** carried out between 1932 and 1972 by the **U.S. Public Health**

African-American men in Alabama under the guise of receiving free health care from the United States government. This criminal scientific study birthed most of the knowledge on the sequence of disease progression in syphilis. This is now documented in most medical books that describe the disease, so it was certainly a study that had an aim. How would it differ from scientists willfully partaking in developing and improving upon existing weapons of mass destruction such as nuclear warheads under the cloak of improving their nation's national security? Is the freezing of prisoners of war in order to study the effects of hypothermia be any different from the above two, no matter what knowledge stood to be obtained thereof?

The imperfections of Science

In the opening chapter, I alluded to the flawed or imperfect nature of science. To understand this, is to grasp the fact that the scientific facts of today will end up being more or less antiquated or even completely discarded, as newer scientific theories emerge. Some of what we consider to be scientific fact today are just the highly cerebral thoughts of men and women endowed with almost superhuman levels of intelligence and reasoning. I will illustrate this using the ideas of a Greek astronomer named Ptolemy.

Ptolemy in the second century developed a model of the solar system with the earth at the center. It was a platform that predicted the positions of heavenly bodies with a degree of accuracy that seemed acceptable. His model was generally adopted at the time with the Earth at the center of the solar system. The irony is that Ptolemy recognized a flawed assumption within his model-he had assumed that the moon followed a path that would occasionally bring it twice as close to the earth as at other times- but nevertheless believed his model was still a correct illustration of the heavens. With current knowledge we can all see how far off from the currently accepted model of the solar system he actually was, from the wrong location of our planet to the wrong assumption concerning the lunar distance from the Earth.

Science is still battling with its inherently flawed nature that tends to throw up inaccurate theories and in other cases incomplete ones. Einstein's general relativity breaks down at the level subatomic matter, where quantum mechanics holds sway. These two theories are considered to be partial theories. The search for a single unified theory that explains all phenomena is still ongoing.

The truth of the matter still remains the fact that science has been a wonderful, albeit imperfect tool that has given us lots of insight and understanding when it comes to explaining the happenings in our universe.

However, I make bold to say that mankind was and is smart enough to have produced much of today's technology without the rigid backbone of science. Observational trial and error, documented over time would have produced sound technologies without necessarily worrying about the intellectually mind bending theories that try to explain every phenomena. The work of many great inventors like Thomas Edison proves this my premise to be an accurate one. His work that produced the first commercially viable electric light bulb was a drawn out series of initial failures. **Wheels were invented** in Mesopotamia **around 3,500 BC, with no science to draw from there are literally thousands of inventions that grew out of this alternative pathway.**

Most theories over time reveal their intrinsic limitations. Even Einstein had to continually tinker with his theory of general relativity, the cosmological constant was one aspect he found particularly unsettling. Some scientists are revisiting the cosmological constant as it might actually have a central place in the search for answers about the cosmos.

The flawed nature of science is captured in this statement taken from the Stephen Hawking and Leonard Mlodinow book, *A Briefer History of*

Time- "Any physical theory is always provisional, in the sense that it is only a hypothesis: you can never prove it. No matter how many times the results of experiments agree with some theory, you can never be sure that the next time a result will not contradict the theory. On the other hand, you can disprove a theory by finding even a single observation that disagrees with the predictions of the theory."

Reference

1. "Science". *Encyclopædia Britannica*. Retrieved July 12, 2016.
2. Michael Ernest Smith and Marilyn A. Masson (2000). *The Ancient Civilizations of Mesoamerica*.
3. Killion, Thomas W., *Gardens of Prehistory: The Archaeology of Settlement Agriculture in Greater Mesoamerica*, University of Alabama Press, 1992.
4. Weiss, E., Mordechai, E., Simchoni, O., Nadel, D., Tschauner, H. (2008). Plant-food preparation area on an Upper Paleolithic brush hut floor at Ohalo II, Israel. Journal of Archaeological Science , 35 (8), 2400-2414.
5. Peter T. Daniels, "The Study of Writing Systems", in *The World's Writing Systems*, ed. Bright and Daniels.
6. Aveni, A.F. (1979). "Astronomy in Ancient Mesoamerica". In E.C. Krupp. *In Search of Ancient Astronomies*. Chatto and Windus.
7. Donald Hill: *A History of Engineering in Classical and Medieval Times*. 263; 8 plates, 52 text figs. London and Sydney: Croom Helm, 1984.
8. Homer (May 1998). *The Odyssey*. Translated by Walter Shewring. Oxford University Press. p. 40.
9. Bynum, W. F.; Hardy, Anne; Jacyna, Stephen; Lawrence, Christopher; Tansey, E.M. (2006). "The Rise of Science in Medicine, 1850–1913". *The Western Medical Tradition: 1800–2000*. Cambridge University Press. pp. 198–199.
10. Pingree, David (1998), "Legacies in Astronomy and Celestial Omens", in Dalley, Stephanie, *The Legacy of Mesopotamia*, Oxford University Press, pp. 125–137,
11. Evans, Jonathan St. B. T.; Newstead, Stephen E.; Byrne, Ruth M. J., eds. (1993). *Human Reasoning: The Psychology of Deduction* (Reprint ed.). Psychology Press.
12. Baird, Forrest E.; Walter Kaufmann (2008). *From Plato to Derrida*. Upper Saddle River, New Jersey: Pearson Prentice Hall.
13. Sarma, SR. *The archaic and the exotic: studies in the history of indian astronomical instruments*. Manohar Publishers, 2008.
14. Keynes, Milo (20 September 2008). "Balancing Newton's Mind: His Singular Behaviour and His Madness of 1692–93".*Notes and Records of the Royal Society of London*. **62** (3): 293. doi:10.1098/rsnr.2007.0025. Retrieved 26 July 2014.
15. Needham, Joseph (1986). *Science and Civilization in China: Volume 3, Mathematics and the Sciences of the Heavens and the Earth*. Taipei: Caves Books Ltd.
16. Saliba, George (1994). *A History of Arabic Astronomy: Planetary Theories During the Golden Age of Islam*. New York University Press.

17. Adamson, Peter (7 July 2016). *Philosophy in the Islamic World: A History of Philosophy Without Any Gaps*. Oxford University Press.

18. Rüegg, Walter: "Foreword. The University as a European Institution", in: *A History of the University in Europe. Vol. 1: Universities in the Middle Ages*, Cambridge University Press, 1992,

19. Cohen, I. Bernard (1976). "The Eighteenth-Century Origins of the Concept of Scientific Revolution". *Journal of the History of Ideas*. **37** (2): 257–288. doi:10.2307/2708824

20. Dreyer, J.L.E. (1953), *A History of Astronomy from Thales to Kepler*, New York, NY: Dover Publications,

21. Finocchiaro, Maurice A. (1989). *The Galileo Affair: A Documentary History*. Berkeley, CA: University of California Press.ISBN 0-520-06662-6.

22. J J O'Connor; E F Robertson (September 1996). "Mathematical discovery of planets". Retrieved 2009-09-11.

23. Boyer, Carl. The History of Calculus. New York: Dover Publications, 1949.

24. Maxwell, James Clerk (1865). "A dynamical theory of the electromagnetic field" (PDF). *Philosophical Transactions of the Royal Society of London*. **155**: 459–512.

25. GW150914: LIGO Detects Gravitational Waves". *Black-holes.org*. Retrieved 18 April 2016.

26. Higgs, Peter (1964). "Broken Symmetries and the Masses of Gauge Bosons". *Physical Review Letters*. **13** (16): 508.

27. Webb Science: The End of the Dark Ages: First Light and Reionization". NASA. Retrieved 9 July 2014.

28. Ptolemaic Astronomy, Islamic Planetary Theory, and Copernicus's Debt to the Maragha School". *Science and Its Times*.Thomson Gale. 2006.

29. Rugh, S, Zinkernagel, H. (2001). "The Quantum Vacuum and the Cosmological Constant Problem". *Studies in History and Philosophy of Modern Physics*. **33** (4): 663–705.

Chapter Four

Glimpses of things less understood

There are things that occur that do not precisely fit into the comfort zones of science and supposed rational thinking.

Before we proceed there are two things I wish to state clearly. First, I am not a fan of the paranormal and secondly, I do not know much about the supernatural either. Having gotten that off my chest, I know there are things that happen that can't readily be explained by science nor by any known constructive thought process that currently exists. So where do these phenomena fit in?

***

What are the chances of getting struck by lightning once in a person's lifetime? It has been estimated that the odds are 1: 10,000 over a period of 80 years. With these slim odds, how does one person get struck by lightning seven odd times and lives through each experience?

A United States park ranger, Roy Cleveland Sullivan, was nicknamed the "Human Lightning Rod" having been hit by lightning seven different times between 1942 and 1977. He survived each strike and is recognized by the Guinness World Records in the above regard. It becomes even more mind-boggling when you consider that each bolt of lightning holds up to several million volts of electricity in it. Does Roy's track record with atmospheric electricity fit into any normal pattern? The answer is an unequivocal no, it certainly does not.

There are even records of Roy fleeing from rain clouds in a bid to avoid his body's apparent affinity for thunder's precursor.

Even if we were to factor in the slightly higher odds, given the fact that his occupation and location (Virginia, USA) slightly increase his chance of getting hit at least once, but seven times? Now, that has got to be anything but definitely not a series of chance events.

From a scientific standpoint it could be argued that Roy might have had a unique body type with a composition that attracts electricity to itself like a natural lightning conductor. Even if this scenario where to be correct, wouldn't others like Roy have similarly frequent hits as well?

It could also be argued that the chances of winning the lottery more than once present even more formidable odds yet people still beat these odds. One example is that of a British couple David and Kathleen Long, who won their first £1m in 2013, then went on to repeat the feat in 2015, beating odds of more than 100 billion to one; supernatural or super lucky? Either way, it's difficult to even try to start thinking up an explanation.

***

The acclaimed boxer Evander Holyfield was a formidable pugilist who had won world championship titles in the cruiserweight and heavyweight divisions. He announced his retirement from the sport in 1996 after a 12 round bout with Michael Moorer. After the fight Holyfield's doctor, Ronald Stevens, described it as being totally inexplicable how Holyfield could have gone through the bout given the fact that he had a congenital heart condition and fought 12 rounds in heart failure. It is important to note that the severity of heart failure is a spectrum that ranges from mild to severe, and it reflects the hearts inability to adequately pump and maintain the blood supply needed by the body. It is completely different from cardiac arrest which is frequently associated with sudden death in severe cases. So was this a supernatural feat or mere hyperbole? We may never know. One thing that is certain is that this information, led to Holyfield's decision to quit the sport rather than face the inevitable prospect of premature demise.

Is it possible to get through 12 rounds of boxing with heart failure? Two important factors that make this remotely possible are 1) the supreme level of conditioning many top athletes have, doesn't require them to fully exert themselves in difficult and physically challenging situations, 2) each round of boxing lasts for a duration of 3 minutes-a good proportion of which is spent being on the defensive and clinging onto your opponent- that is followed by an interlude of sixty seconds for the fighters to rest. This period of rest would be a lifesaver for a boxer with even mild heart failure.

Holyfield would later on have an encounter with a popular televangelist at a revival meeting where, after recounting his career ending heart condition, he was told on live TV by the evangelist that: "You are healed." Holyfield assured him that was impossible, but he was simply told to go back to his doctors to get retested.

He did just that, and to his and the doctor's surprise, all the tests came up negative.

When he applied to have his boxing license reinstated, the Nevada Athletic Commission had their doubts. So they sent him to have a comprehensive evaluation at the renowned Mayo Clinic. After a week-long series of testing, the physicians here pronounced Holyfield's heart 100% healthy and fit.

Interestingly he returned to boxing the next year and to the amazement of all he would later go on to beat the likes of Mike Tyson (he actually beat Tyson twice) after his comeback.

Now could someone give me a clear cut way of explaining what happened to Evander Holyfield? A misdiagnosis of the initial heart condition, divine healing, the heart somehow went into auto-repair mode, or his body threw up stem cells that built a new heart?

A misdiagnosis was most unlikely as the diagnostic protocol is quite straight forward for heart failure. The other possibilities mentioned cannot be ignored as each represents a pathway to healing a heart in failure.

This reminds me of an interesting article I once read in Time magazine, that explored the concept of miracles, but the article as unbalanced as it was in trying to "not acknowledge" the existence of events that defy the realm of scientific explanation, it did however point out the case of a little girl who had an inoperable brain tumor and was given a short time to live.

Well her parents sought a different kind of remedy and had their church elders pray for the young damsel. By her next hospital appointment there was no trace of the tumor. The doctors labeled it a case of spontaneous resolution.

Don't similar questions pop up here as in the case of Evander Holyfield, what do you think?

***

The field of near death experience (NDE) is quite intriguing. It describes a personal experience that is closely associated with impending death. It includes an array of descriptions ranging from a sense of detachment from the body, warmth, security, feelings of levitation, the presence of light, seeing a tunnel of light, being aware of voices and activities around the person while unconscious.

The similarities between NDEs across various cultures is pointer to the fact that it is not based on frivolous claims.

Near death studies draw from neuroscience, psychiatry and other medical disciplines.

There was a particular study on NDEs that caught my attention.

I will try to describe its basic highlights without being too technical, the study was carried out at Southampton General Hospital and it involved 63 individuals who had all survived cardiac arrest. They all had reached a point where they had no pulse, no respiration, and pupils that where fixed and dilated- parameters that point to the cessation of brain function.

Dr. Sam Parnia, a clinical research fellow and co-author of the study is of the opinion that the rapid loss of brainstem activity during cardiac arrest should make it impossible to sustain lucid processes or form lasting memories.

Out of the 63 survivors only seven of them recollected emotions and visions that are typical of NDEs. A clear pointer to the fact that this may not be a universal experience associated with dying or being close to death.

According to Dr. Parnia, the recollections, unlike hallucinations, were "highly structured, narrative, easily recalled and clear."

Opinions are strongly divided within the scientific community as to the actual explanation for these NDEs. Some say it clearly shows the continuity of consciousness after the brain shuts down and life approaches its known end. Others resist this notion and have put forward various theories to place this event within the confines of terminal brain activity, such as a final surge in neuronal activity in the brain.

Some point to the fact that phenomena such as out of body experiences are also now known to occur during interrupted sleep patterns that immediately precede sleeping or waking. For instance, sleep paralysis, or the experience of feeling paralyzed while still aware of the outside world, is reported in up to 40 percent of all people and is linked with vivid dreamlike hallucinations that can include the sensation of floating above one's body. A 2005 study found that out of body experiences can be artificially triggered by stimulating an area of the brain called the right temporoparietal junction, suggesting that confusion regarding sensory information can radically alter how one experiences their body.

My problem with the theories is that most fail to explain the common features that are seen in NDEs, irrespective of the study population.

A counter argument would be that certain chemicals induce similar hallucinations irrespective of individual's life experiences, a strong argument but not essentially watertight. When such hallucinations are analyzed in great detail, many apparently subtle differences become more pronounced; eventually showing that few if any have totally identical hallucinations.

NDEs can also vary quite widely between people of different cultural backgrounds. Some suggest it is the low levels of oxygen in brain tissue that provoke these experiences. This argument looks plausible but how do we explain the occurrence of spontaneous hallucinations amongst a significant proportion of mentally and physically healthy individuals?

Some of the science that tries to explain NDEs makes sense but there are still loopholes. For instance, why would a person from California USA have a nearly identical NDE as a monk in Tibet, even though they have few if any shared common life experiences?

Why do we have people who speak the same language with different accents? English is spoken with such diverse accents it is amazing. Even within England the accents with which English is spoken varies quite significantly. Before we try to explain this let us examine a rare medical condition called *Foreign Accent Syndrome* (FAS), here patients develop speech patterns that are perceived as a foreign accent. This usually results from a stroke, head trauma or migraines. One case that was well documented was that of a Norwegian lady named Astrid who in 1941 was hit in the head with shrapnel during a raid during World War II. She would later see neurologist George Herman Monrad-Krohn who was quick to notice that despite her fluent Norwegian that she had: "such a decided foreign accent that I took her for German or French," he wrote.

Quite recently, Lyndsey Nickels, a professor of cognitive science at Macquarie University, put forward an explanation for this condition using a widely covered case in which an Australian, Leanne Rowe developed a French accent following a car accident. According to Lyndsey "vowels are very susceptible," to being distorted by slight differences in tongue, lip or jaw placement. "Different languages have different vowels, and within a language one of the main differences between accents is in the vowels." It may be that the speech changes brought on by brain injury are just familiar enough to remind us of accents.

To say that such an explanation is just too simplistic is to state the obvious. By extrapolation, do accents arise from brain trauma in normal people? Of course not! That damage occurs in the brain prior to the development of this syndrome is not in doubt but why does it occur in such a small percentage of cases of stroke or head trauma victims even when brain imaging clearly reveals damage to areas of the brain responsible for speech, so why isn't FAS more commonly seen?

It is my opinion that this is an area that needs to be researched to a greater depth, no expert should just conjure a commonsense explanation that lacks the backing of concrete scientific facts such as neurotransmitter variations, actual brain imaging that is reproducible in other patients, as well as ruling out the possibility of prior exposure to the new accent. Without these any attempt to explain FAS would amount to no more than intellectual bluster and guess work.

***

On a more personal level, I will share a few true life experiences that I have had. I have carefully chosen just two events.

The first being when I had what I will call a vision, this is for the lack of a better term to use, several years ago while I was working in a foreign country.

To understand why I am even discussing this, you need to realize that it is the fact that I either rarely have dreams or I absolutely forget their details when I awake, that made this experience all the more eerie.

So you can imagine me waking up in a cold sweat by 2.12am on a winter morning, after having a very life like dream.

Don't you worry, there were no issues with the heating of my bedroom. I had just had a very vivid and detailed (dream, vision, revelation, please call it whatever you want) account of how I was going to lose my job while I had been fast asleep.

When was it going to happen? Now that's the weird part, it was going to happen that same day. I remembered every single detail of the dream including the date on the newspaper I was supposedly reading on the train to work.

Just to prove to myself that this was all some kind of coincidence, I gathered my wits about myself and managed to get another one hour or so of shut eye.

I got up with a sense of "I don't believe in superstition" bravado.

I proceeded to take a nice hot shower, get dressed, then grab the train to work. I read the same newspaper I had in the dream with the same front page story (it felt very awkward but I was a man of scientific leanings, right?). The dream seemed to be unfolding with every fabric of it seemingly falling into place.

I later wished that I had been a tad superstitious, because unfortunately everything played out the way I had dreamt it.

Now, kindly explain how that is even vaguely scientifically possible?

The second event happened when my wife was pregnant with our first child, she was about 6 weeks down the road, and the first person we told was Ifie (not her real name). Ifie was a first degree relative, the type you feel at ease with when it comes to sharing private matters.

She smiled and chuckled happily for a few seconds, then she looked me straight in the eye and told me that she already knew that my wife was expecting. As if that wasn't a big enough surprise, she said the child would be a light skinned boy. At this point I couldn't help but laugh derisively. I was like, "Yeah right!"

We hadn't even had the first scan when this conversation took place. To make this sound even more outrageous, I happen to be dark in complexion while my wife is much lighter than myself. The child in her womb could easily have any skin tone that falls anywhere on the skin color spectrum. So I dismissed it as just the wishful thinking of someone else.

I remember carrying my son for the first time, it was that kind of tender moment you never forget. He was indeed light skinned alright. Ifie's words haunted the deepest recesses of my mind, did she really have some abstract power of clairvoyance or was she just great at guessing?

I will leave you to be the judge of that. That wasn't the first, neither was it the last time she had displayed this uncanny ability to predict future events accurately, she was somewhat of a local legend. Before my personal one on one encounter, I used to dismiss it as basal superstitious thought patterns.

Historical documents are replete with incidences like the ones I have mentioned where people can state things they have next to no firsthand knowledge of with profound accuracy.

One quick example is that involving a psychic, Annette Martin. She was able to help police locate the missing body of a retired US paratrooper. A certain Dennis Prado had gone missing from his apartment and police had been unsuccessful in locating his whereabouts. With no further leads, the chief investigating officer, a sergeant with the Pacifica (California) Police, opted to contact a psychic detective.

Annette was given a map, she then proceeded to circle a small spot on the map, the size of which was no bigger than two city blocks. The only problem was that the area had already been searched before and Prado had not been found, an extremely strange coincidence I must say.

A search and rescue officer commenced a fresh search with the help of a search dog, as Martin suggested: "A search dog is going to find him." They found the body covered with dirt at the same location, as Martin had indicated.

Why this case stands out is because of the abysmal record psychics have developed over the years while working with various police departments.

So the real question here is: are there people with the power of clairvoyance or who actually see visions?

Where do premonitions fit into in this untidy mix? But then again I am a man of scientific leanings so I try to analyze these from an evidence based standpoint which unfortunately doesn't yield consistent answers?

The only certain thing seems to be that not all physical phenomena are easily figured out using science or logic.

***

Now let me digress a little and discuss a strange paranormal phenomena that has been documented amongst the Australian aborigines. The phenomenon of "bone pointing" by ritual executioners called the *kurdaitcha*.

This practice amongst the indigenous aboriginal populace is documented and known to produce the death of its victim.

Such victims of bone pointing tend to die within days of having been "cursed".

The event is well recognized and medical staff have been trained to handle the effects of illness arising from bone pointing.

There are cases of victims who were hospitalized and extensive laboratory investigations failed to turn up any abnormalities, despite the fact that these victims were obviously dying.

A good question at this point is, "Was some otherworldly force at play here or is there a simpler explanation for this?"

In medical science these terminal effects are suggested to be psychosomatic. Basically the mind is so convinced of the likelihood of death that the body follows the prompting of the mind to die.

As a doctor, I have seen psychosomatic disorders at play in many patients. They are real disorders some with pronounced debilitating effects but to put the death of victims of the "bone pointing syndrome" down to psychosomatic causes doesn't sit very comfortably with me. The human psyche is far too complex for one event to consistently produce a particular outcome, death in this case.

To further follow my train of thought let us examine a condition that has psychosomatic underpinnings- panic disorder. This is a condition where fear (an evolutionary protective mechanism or so it is thought to be) is triggered without any obvious external evidence of danger in sufferers. Like a faulty alarm clock their bodies manifest the responses seen in a normal individual confronted by danger. The so called "fight or flight" response that is caused by a release of adrenaline, is set in motion. Over time this faulty mechanism self-perpetuates itself leaving the sufferer in a state of recurring cycles of irrational panic, behaviors that promote avoidance of triggers and a catastrophic outlook. No matter how recurrent or ingrained these catastrophic thoughts of impending demise are, they never truly play themselves out.

So in both cases, the mind believes albeit irrationally, that something catastrophic is lurking round the corner. Why then is it that in only one group their bodies actually act out this conceived notion? Genetic makeup, willpower, mental conditioning and resilience, might in play a role here. These factors do not in any way invalidate the possibility of undefined factor causing these deaths, psychosomatization alone wouldn't consistently produce the predictable outcome of death in bone pointing syndrome.

***

In a nutshell, the various events mentioned may seem not to have any direct link. You might even be tempted to ask "Where exactly are we going with all this information?" I chose unrelated paranormal, supernatural and freakish events to demonstrate that there is more to life on earth than the physical things the eyes can behold.

My conclusion is that our world extends beyond the physical, there is obviously an aspect of our existence that is unseen. I will simply term it the realm beyond sensory appreciation. Certain people seem to be able to tap into this realm while others are oblivious to its existence.

That you cannot see gravity doesn't in anyway alter the fact that it constantly exerts its influence on you and the world around you.

Now, you are free to draw your own conclusions from the above stories as well. As my views are in no way sealed with finality.

Reference

1. John Friedman (2008). *Out of the Blue: A History of Lightning: Science, Superstition, and Amazing Stories of Survival*. Delacorte Press.
2. Campbell, Ken (2000). *Guinness World Records 2001*. Guinness World Record Ltd. p. 36.
3. Parnia S, Waller DG, Yeates R, Fenwick P. "A qualitative and quantitative study of the incidence, features and aetiology of near death experiences in cardiac arrest survivors". *Resuscitation*. Feb; 48 (2):149-56, 2001.
4. Augustine, K. "Psychophysiological and cultural correlates undermining a survivalist interpretation of near-death experiences". *Journal of Near Death Studies*, 26 (2):89-125 (2007).
5. Olaf Blanke and ShaharArzy (2005). "The Out-of-Body Experience: Disturbed Self-Processing at the Temporo-Parietal Juncti". *The Neuroscientist*. **11** (1): 16–24. doi:10.1177/1073858404270885.
6. Kurowski, K.M.; Blumstein, S.E.; Alexander, M. (1996). "The foreign accent syndrome: a reconsideration" (PDF). *Brain and Language* (54(1)): 1–25.
7. Monrad-Krohn, G. H. "Dysprosody or Altered 'Melody of Language'." Brain 70 (1947): 405-15.
8. Lyndsey Nickels. Explainer: what is foreign accent syndrome? www.theconveration.com. Accessed on 08-06-2016.

CHAPTER Five

Unanswered question: Where did life come from?

The origin of life on Earth has been and remains a topic in which no concrete consensus has been reached within or outside the scientific community. It is an unresolved scientific problem with a plethora of ideas but few precise facts. The leading theories amongst geochemists, evolutionists and advocates of intelligent design vary both in outlook and conclusions.

The science of life's beginnings.

Based on scientific extrapolation our planet, the Earth is 4.6 billion years old. The emergence of life began quite early in its timeline. Life on our planet appeared sometime between 4.4 billion years ago when liquid water first appeared and 3.5 billion years ago when the earliest undisputed evidence of life existed during the Eoarchean Era. This evidence of early life was discovered in **microbial mat fossils** that were found in 3.48 billion year old mat fossils discovered in Western Australia. A microbial mat is a multilateral sheet of microorganisms, mainly bacteria and archaea, these mats are the earliest form of life on Earth for which there is good fossil evidence. Other discoveries include biogenic graphite in meta-sedimentary rocks that are 3.7 billion years old in southwestern Greenland. More recently, University of California, Los Angeles (UCLA) geochemists found strong evidence that life likely existed on Earth at least 4.1 billion years ago- 300 million years earlier than previous research suggested. These researchers at UCLA, studied over ten thousand zircons originally formed from molten rocks or magma from Western Australia. Zircons are heavy, durable minerals related to the synthetic cubic zirconium used for imitation diamonds. Their ability to capture and preserve their immediate environment means that they can serve as time capsules.

The team of scientists identified 656 zircons containing dark specks that could possibly contain evidence of early life then closely analyzed 79 of them with Raman spectroscopy, a technique that reveals the molecular and chemical structure of ancient microorganisms in three dimensions. One of the 79 contained graphite- pure carbon- in two locations. The researchers noted that the graphite is older than the zircon containing it. The zircon is 4.1 billion years old, based on its ratio of uranium to lead; they however do not know how much older the graphite is.

Abiogenesis

Abiogenesis or biopoiesis is the process of living matter evolving from self-replicating nonliving molecules. This hypothesis began to gather momentum by 1924 when Alexander Oparin alluded to the possibility that atmospheric oxygen prevented organic molecule synthesis. Oparin in his hypothesis; *the Origin of Life*, argued that a "primeval soup" of organic molecules could be created in an oxygen free atmosphere through the activity of sunlight. These would then go on to interact in ever more complex fashions forming droplets. These droplets would fuse with other droplets. Reproduction in a very primitive form would occur through fission into daughter droplets, and so have a primordial **metabolism** in which those factors that promote "cell integrity" survive, while those that do not become extinct. Organic molecules are the

necessary building blocks for the evolution of life and Oparin hypothesized on the possible pathway for their generation.
Many modern theories of the origin of life still take Oparin's ideas as a starting point.
At roughly the same time that Oparin's work surfaced, it was also suggested by **J.S Haldane,** that the Earth's pre-biotic oceans would have formed a "hot dilute soup". In this soup, organic compounds, the building blocks of life, could have formed.

The study of abiogenesis utilizes three main types of considerations: 1)the **geophysical**, 2)the **chemical**, and 3)the biological. More recent approaches have attempted a combination of all three of these. **This theory of abiogenesis has been rigorously studied using both laboratory experiments as well as drawing conclusions from the genetic information of modern organisms. It aims at making reasonable conjectures about what pre-life chemical reactions may have given rise to living systems.**

Here it is believed that all life forms we know of today evolved from a common primitive ancestor referred to as the last universal common ancestor, LUCA, that existed 3.6 billion years ago. Prior to the appearance of LUCA there is uncertainty as to which of these came first; metabolism or genetics? The hypothesis supporting genetics first is called the RNA world hypothesis, and the one which backs metabolism first is called the Protein world hypothesis. However it is widely accepted that irrespective of which came first, that current life descended from an RNA world. This factors in the possibility that RNA based life may not have been the first type of life to have existed. The common link between most of the work on the origin of life using the abiogenesis model was the idea that before biological evolution began there must have been a process of chemical evolution. The Miller-Urey experiment and others like it show that amino acids can be synthesized from inorganic compounds in conditions that mimic those that existed on early Earth. The Miller–Urey experiment was a chemical **experiment** that simulated the conditions thought at the time to be present on the planet during its early years, and tested the chemical origin of life under those conditions. The experiment used **water**, **methane**, ammonia, and hydrogen. The chemicals were all sealed inside a sterile 5 liter glass flask connected to a 500 ml flask half-full of liquid water. The liquid water in the smaller flask was heated to induce **evaporation**, and the water vapor was allowed to enter the larger flask. Continuous electrical sparks were fired between the electrodes- within the set up- to simulate **lightning** in the water vapor and gaseous mixture, and then the simulated atmosphere was cooled again so that the water condensed and trickled into a U-shaped trap at the bottom of the apparatus. At the end of one week of continuous operation, the boiling flask was removed, and mercuric chloride was added to prevent microbial contamination. The reaction was stopped and evaporated to remove impurities. Using paper chromatography, Miller identified five amino acids present in the solution. Intriguingly after Miller's death in 2007, scientists examining sealed vials preserved from the original experiments were able to demonstrate that there were well over 20 different amino acids produced in Miller's original experiments.

More recent studies by chemists reveal that a pair of simple compounds- acetylene and formaldehyde, which would have been abundant on early Earth, can give rise to a network of simple reactions that produce the three major classes of biomolecules - nucleic acids, amino acids, and lipids- needed for the earliest form of life to get its start. It is essential to point out that these new insights do not prove how life actually started. They primarily help provide explanations for one of the confounding mysteries in modern science, how life on Earth came into existence. The RNA world hypothesis was greatly bolstered in 2009, when chemists led by John Sutherland at the University of Cambridge in the United Kingdom reported that they had discovered, that the relatively simple precursor compounds of acetylene and formaldehyde could undergo a sequence of reactions to produce two of RNA's four nucleotide building blocks, demonstrating a plausible pathway to how RNA could have formed independent of the need for enzymes in the primordial soup. Because of the relative complexity of acetylene and formaldehyde, this threw up a new question as to where these came from. Sutherland and his colleagues set out to work backward from those chemicals to see if they could find a route to RNA from even simpler starting materials. They were eventually able to, the team created nucleic acid precursors starting with just hydrogen sulfide- H_2S, hydrogen cyanide- HCN, and ultraviolet (UV) light. The team pointed out that the conditions that produce nucleic acid precursors also create the starting materials needed to make natural amino acids and lipids. That suggests a single set of reactions could have given rise to most of life's building blocks simultaneously. The conditions on early Earth were a favorable setting for those postulated reactions. Hydrogen cyanide is abundant in comets, which rained down steadily for nearly the first several hundred million years of Earth's history. This happened between 3.8 and 4.1 billion years ago, a time known as the **Late Heavy Bombardment.**The impacts would also have produced enough energy to synthesize HCN from hydrogen, carbon, and nitrogen. Likewise, H_2S was thought to have been common on early Earth, as was the UV radiation that could drive the reactions.

Deep sea vent hypothesis

This hypothesis falls within the overall scope of abiogenesis.
A hydrothermal vent is a fissure in a planet's surface from which geothermally heated water escapes. Hydrothermal vents are commonly

found close to places that are volcanically active, areas where tectonic plates are moving apart, ocean basins, and hot-spots. Common land types include hot springs, **fumaroles** and **geysers**. Under the sea, hydrothermal vents may form features called black smokers- having a characteristic black color due to the iron and sulfide content that precipitates from its super heated water. The deep sea vent or alkaline **hydrothermal vent** theory for the origin of life on Earth posits that life may have taken off within submarine hydrothermal vents.William Martin and Michael Russell have suggested "that life evolved in structured iron monosulphide precipitates in hydrothermal seepage-sites that had a favorable redox, pH, and thermal gradient between sulphide-rich hydrothermal fluid and iron-containing waters of the early ocean floor. The naturally arising, three-dimensional compartmentation observed within fossilized seepage-site metal sulphide precipitates, indicates that these inorganic compartments were the precursors of cell walls and membranes found in free living unicellular microbes called prokaryotes." Michael Russell demonstrated that alkaline vents created an abiogenic chemiosmotic gradient, in which conditions are ideal for an abiogenic hatchery for life.

Jack W. Szostak suggested that geothermal activity provides greater opportunities for the origination of life in open lakes, where there is a buildup of minerals.

In 2010, based on spectral analysis of sea and hot mineral water, Ignat Ignatov and Oleg Mosin demonstrated that life may have predominantly originated in hot mineral water. Hot mineral water that was rich in **bicarbonate** and calcium ions had the optimal range for this to happen. This case is similar to the origin of life in hydrothermal vents, but with bicarbonate and calcium ions in hot water.

An environment akin to these is where stromatolites have been created. A *stromatolite* is a solid layered structure created by single-celled microbes called cyanobacteria (blue-green algae). David Ward of **Montana State University** described the formation of stromatolites in hot mineral water at the **Yellowstone National Park**. Stromatolites survive in hot mineral water and in proximity to areas with volcanic activity. In 2011, Tadashi Sugawara from the **University of Tokyo** created a protocell in hot water. A protocell is a self-organized, endogenously ordered, spherical collection of lipids proposed as a stepping-stone to the origin of life.

Experimental research and computer modeling suggest that the surfaces of mineral particles inside hydrothermal vents have catalytic properties similar to those of enzymes and are able to create simple organic molecules, such as **methanol** and pyruvic acid out of the dissolved carbon dioxide in the water.

Panspermia

The Panspermia hypothesis suggests that microscopic life was spread out throughout the universe by meteoroids, asteroids, comets and planetoids. It is postulated that shortly after the Big Bang happened 13.8 billion years ago, that the biochemistry of life began during a habitable epoch when the universe was still an infantile 10-17 million years old.

Panspermia as a hypothesis puts forward the likelihood of microscopic life forms, such as extremophiles, being able to survive the effects of space while they are transported within celestial debris. Extremophiles are microorganisms that thrive in **extremely hostile environments** including intensely hot, highly acidic environments, under extreme pressure and extremely cold places; previously believed to be inhospitable to any form of life. These organisms may be dormant as they travel in this manner through space for extended periods of time before colliding at random with planets or interfusing with protoplanetary disks. Where and if the right conditions exist on the new planet's surface the organisms become active and the process of evolution starts.

The concept of panspermia was articulated by the Swedish chemist Svante Arrhenius in the early 1900s. The duo of Fred Hoyle and Chandra Wickramasinghe were influential proponents of panspermia. They proposed that some of the dust in interstellar space was largely organic. This was later proved correct by Wichramasinghe. **It was Wickramasinghe who demonstrated that there are complex organic polymers, in space.** These molecules are closely related to cellulose, which is very abundant in biology.

If the building blocks of organic life are located in cosmic dust is it not a reasonable extrapolation that simple life forms can develop in interstellar or intergalactic space? I would certainly not rule out this possibility particularly in the early hours that followed the Big Bang.

During a presentation in April 2009, renowned physicist Stephen Hawking gave his opinion on the possibility of life in other parts of the universe. He said "Life could spread from planet to planet or from stellar system to stellar system, carried on meteors."

Many microbes have been shown to exhibit remarkable levels of resilience when exposed to extreme conditions, for instance recent research done on extremophiles in Japan showed that a certain bacterial organism *Paracoccus denitrificans* could be cultivated and exhibit robust cellular growth within an environment with the equivalent of over 400,000 times the gravity on earth. Such conditions are nearly exclusively found in cosmic environments such as very massive stars or in the shock waves of a supernova. This research has positive implications for the theoretical feasibility of panspermia.

Several studies and simulations done in low Earth orbit and in laboratories point to the ability that some simple organisms have, been able to survive the rigors of ejection, entry and impact with atmospheric and terrestrial planetary components.

The EXPOSE experiments- which were well cataloged astrobiology experiments conducted outside the International space station between 2008 and 2015, these tests exposed an extensive array of biomolecules, spores and microorganisms to the vacuum of space and solar flux for about a year and a half. Some organisms survived in a dormant state for prolonged lengths of time.

Panspermia whether it is a correct or incorrect hypothesis only attempts to point to where life came from but not how life actually arose.

Intelligent Design

Intelligent design (ID) is the view that it is possible to infer from empirical evidence that "certain features of the universe and of living things are best explained by an intelligent cause, not an undirected process such as natural selection." Intelligent design cannot be inferred from complexity alone, since complex patterns often happen by chance. ID focuses on just those sorts of complex patterns that in human experience are produced by a mind that conceives and executes a plan. According to adherents, intelligent design can be detected in the natural laws and structure of the cosmos; it also can be detected in at least some features of **living things**.

One thing ID theorists emphasize is that it is not an argument from ignorance. Intelligent design is not to be inferred on the simplistic basis that the cause of something is unknown. ID does not claim that design must be without defect; something may be intelligently designed albeit flawed, as are many human made items.

ID may be considered to consist only of the minimal assertion that it is possible to infer from empirical evidence that some features of the natural world are best explained by factoring in an intelligent agent. It conflicts with views claiming that there is no real design in the cosmos (e.g., materialistic philosophy) or in living things (e.g., Darwinian evolution) or that design, though real, is undetectable (e.g., some forms of theistic evolution). Because of such conflicts, ID has generated considerable controversy.

Historical background of Intelligent Design

Inferring design from nature or natural phenomena dates back to the time of Plato and Aristotle, and Christian writers have also used the inference for centuries to argue for God's existence and attributes. The term intelligent design appeared in the 1980s. In 1982, the cosmologist Fred Hoyle used the term while writing that unless a person is "deflected by a fear of incurring the wrath of scientific opinion, one arrives at the conclusion that bio-materials with their amazing measure of order must be the outcome of intelligent design." Not too long afterward, the chemist Michael Polanyi argued that the information in DNA could not be reduced to physics and chemistry, insisting that something more was needed. Charles B. Thaxton, a chemist himself was intrigued by Polanyi's views. Thaxton later explained his preference for the term intelligent design over the concept of creationism.

In 1984, he teamed up with materials scientist Walter L. Bradley and geochemist Roger Olsen to publish *The Mystery of Life's Origin,* which criticized chemical evolution. The idea that unguided natural processes produced the first living cells abiotically from non-living materials. The authors distinguished between order- such as found in crystals, complexity- such as found in random mixtures of molecules, and specified complexity- the information rich complexity in biological molecules such as DNA. Relying on the uniformitarian principle "that the kinds of causes we observe producing certain effects today can be counted on to have produced similar effects in the past," the authors argued, "What is needed is to identify in the present an abiotic cause of specified complexity." Thaxton, Bradley, and Olsen concluded: "We have observational evidence in the present that intelligent investigators can and do build contrivances to channel energy down nonrandom chemical pathways to bring about some complex chemical synthesis, even gene building. May not the principle of uniformity then be used in a broader frame of consideration to suggest that DNA had an intelligent cause at the beginning?"

In 1991, Berkeley law professor Phillip E. Johnson published *Darwin On Trial,* which critically analyzed the logic and assumptions Darwinists use to rule out design in living things. Johnson concluded: "Darwinist scientists believe that the cosmos is a closed system of material causes and effects, and they believe that science must be able to provide a naturalistic explanation for the wonders of biology that appear to have been designed for a purpose. Without assuming those beliefs they could not deduce that common ancestors once existed for all the major groups of the biological world, or that random mutations and natural selection can substitute for an intelligent designer."

At a conference held in Kunming, China, in 1999, American, European and Chinese scientists discussed the implications of fossils that had been found at nearby Chengjiang. The fossils documented in great detail the abrupt appearance of most major animal body plans (phyla) in the Cambrian Explosion, a feature of the fossil record that gives the appearance of conflict with the branching-tree pattern expected from the Darwinian theory. With such overwhelmingly indisputable fossilized evidence the Darwinian model begins to not only look inaccurate but out-rightly untenable in the realm of evidence based science.

Ideas from prominent ID proponents

In *The Origin of Species,* Charles Darwin wrote: "If it could be demonstrated that any complex organ existed which could not possibly have been formed by numerous, successive, slight modifications, my theory would absolutely break down." In his 1996 book *Darwin's Black Box,* biochemist Michael J. Behe wrote: "What type of biological system could not be formed by numerous successive, slight modifications? Well, for starters, a system that is irreducibly complex. By irreducibly complex I mean a single system composed of several well-matched interacting parts that contribute to the basic function, wherein the removal of any one of the parts causes the system to effectively cease functioning."

Behe described several features of living cells—features unknown to Darwin—that he considered being irreducibly complex. These include the light-sensing mechanism in eyes, the human blood-clotting system, and the bacterial flagellum.

He searched the scientific literature but found no articles proposing detailed, testable explanations of how these and other irreducibly complex systems originated through Darwinian evolution.

Behe argued that biochemists know what it takes to build irreducibly complex systems; it takes design. He wrote: "The conclusion of intelligent design flows naturally from the data itself—not from sacred books or sectarian beliefs. Inferring that biochemical systems were designed by an intelligent agent is a humdrum process that requires no new principles of logic or science. It comes simply from the hard work that biochemistry has done over the past forty years, combined with consideration of the way in which we reach conclusions of design every day."

Although most ID arguments currently focus on design in living things, some focus on design in the cosmos. In *The Privileged Planet: How Our Place in the Cosmos is Designed for Discovery* (2004), astronomer Guillermo Gonzalez and philosopher Jay W. Richards argued that the universe and our place in it are designed not only for life, but also for science.

The authors reemphasize a point made by others—that multiple universal constants, including the strength of gravity, the strength of the electromagnetic force, and the ratio of the masses of the proton and electron, are remarkably fine-tuned for life. Even miniscule alterations to these constants would make the universe uninhabitable. Gonzalez and Richards also point out that the Milky Way is just the right kind of galaxy to support life, and our solar system is situated in a relatively narrow "galactic habitable zone" in the Milky Way that minimizes threats from dangerous radiation and comet impacts, and also ensures the availability of heavy elements needed to form large rocky planets. Our Sun is just the right size and has the necessary stability to support life. Furthermore, the Earth is just the right size to hold an atmosphere, it consists of dry land as well as oceans, and produces a protective magnetic field. Finally, the Moon is just the right size and distance from the Earth to stabilize the tilt of the latter and thereby prevent wild fluctuations in temperature.

Not only is the Earth especially suited for life, but it is also well situated for scientific discovery. Because the Milky Way is a spiral galaxy, it is relatively flat, so that from our vantage point midway from its center to its edge we can enjoy clear views of distant galaxies and the subtle cosmic background microwave radiation that provided evidence for the Big Bang. Our solar system is also well suited to scientific discovery. The simple orbits of the planets, and the large Moon orbiting the Earth, have guided scientists to an accurate understanding of gravity. The same parameters also make possible total solar eclipses, which have played a crucial role in astronomy. During a total solar eclipse, the

Moon exactly covers the face of the Sun, leaving only its tenuous outer atmosphere visible from the Earth. Studying that outer atmosphere has enabled astronomers to make discoveries about the composition of the Sun and other stars. Total solar eclipses have also provided tests of Einstein's theory of general relativity. If the Moon were smaller or larger, or closer or farther away, such discoveries and tests would have been delayed, perhaps indefinitely. To Gonzalez and Richards, it seems as though the size and orbit of the Moon were tailor-made for science.

So the most habitable places in the universe are also the best places to make scientific discoveries about it. According to Gonzalez and Richards: "There's no obvious reason to assume that the very same rare properties that allow for our existence would also provide the best overall setting to make discoveries about the world around us. We don't think this is merely coincidental. It cries out for another explanation, an explanation that suggests there's more to the cosmos than we have been willing to entertain or even imagine." They conclude that the correlation between the factors needed for complex life and the factors needed to do science "forms a meaningful pattern" that "points to purpose and intelligent design in the cosmos."

Astronomer Sir Fred Hoyle a proponent of panspermia, put forward assertions that completely rejected abiogenesis while inadvertently giving credence to ID even though he was an atheist. He stated in a varying array of ways the extreme improbability of life forming spontaneously, or even getting a single functional biological polymer (biopolymer) such as a protein to arise randomly. Hoyle believed that the odds that 10^{50} blind people- each with a scrambled Rubik cube would be able to simultaneously solve it was next to impossible. If this impossible scenario did however occur, then the chance of arriving by random shuffling of just one of the many biopolymers on which life depends might be possible. He considered the dual notions that: 1) the biopolymers needed for living cells and 2) the operating program of a living cell could be arrived at by chance in a primordial soup here on earth would not just be possible. The very notion of this even being the subject of debate was considered by him to be utter nonsense of a high order. Life must plainly be a cosmic phenomenon. In his own words: "The likelihood of the formation of life from inanimate matter is one to a number with 40,000 zeroes after it … It is big enough to bury Darwin and the whole theory of evolution. There was no primeval soup, neither on this planet nor any other, and if the beginnings of life were not random, they must therefore have been the product of purposeful intelligence." Proponents of ID point to analyses such as the ones made by Fred Hoyle as being pointers to the validity of the concept of ID.

In June 2016, theoretical physicist Michio Kaku made a statement that gave a boost to proponents of ID. He opined that the universe was created by "an intelligence."

Criticism of Intelligent design

Critics of ID also point out that the consensus of scientific opinion overwhelmingly favors Darwinian evolution and rejects intelligent design. Many scientific societies in the U.S have issued statements to this effect. ID proponents counter that what matters in science is evidence, not opinion polls, and that history shows that the scientific consensus is often unreliable.

Other critics reject the notion of ID claiming that it can never be scientifically fruitful, because instead of exploring possible mechanisms it merely puts a stop to inquiry by saying "God did it." ID theorists disagree, predicting that scientists who regard living things as designed will discover mechanisms that have been overlooked by scientists who regard living things as accidental byproducts of unguided natural processes.

Creationism

Creationism is the religious belief that the Universe and life originated from specific acts of **divine creation**. There are varied views amongst creationists, it must be said that there is heterogeneity within their ranks.

The account of creation in the book of Genesis presents a picture of the world being created in six days. Many have suggested that this should not be taken literally as the concept of time is different when referring to an ageless being in this case God. They refer to the scripture 2 Peter 3: 8 (the King James Version) which states, "But, beloved, be not ignorant of this one thing, that one day is with the Lord as a thousand years, and a thousand years as one day." Thus they refer to this scripture as an important pointer to their allusion that the concept of time is different for man with regard to his Creator.

The biblical book of Hebrews in chapter 11 verse 3 states "Through faith we understand that the worlds were framed by the word of God, so that things which are seen were not made of things which do appear." This is referenced by some creationist as being quite similar to the scientific model of what existed prior to the Big Bang when the universe as we know it was nonexistent. They say it clearly demonstrates a starting point for the cosmos.

The scientific community mostly reject the idea of creationism, however not all scientists completely reject it. A few even allude to the existence of a Creator. For instance, Albert Einstein said he believed in the pantheistic God of Baruch Spinoza, but not in a **personal god**, a belief he criticized. He also called himself an agnostic, while disassociating himself from the label **atheist**, preferring, he said, "an attitude of humility corresponding to the weakness of our intellectual understanding of nature and of our own being".

Pantheism is the belief that all of reality is identical with **divinity**, or that everything composes an all encompassing, **immanent** god. Pantheism was formalized as a separate theology and philosophy based on the work of the 17th century philosopher Baruch Spinoza.

Physicist Stephen Hawking, a self proclaimed atheist, shocked the scientific community with his recent comments on the plausibility of intelligent design. He proposed **that a "God-like force" played a role in the creation of our Universe, approximately 13.8 billion years ago.**

The truth remains that the actual events that led up to the emergence of life on Earth is still an unresolved matter.

Reference

1. *The New Oxford Dictionary Of English*. Oxford: Clarendon Press. 1998. p. 1341.
2. Schopf J.W. *et al* 2002. "Laser--Raman imagery of Earth's earliest fossils". *Nature* 416 (6876): 73–6.doi:10.1038/416073a.
3. Mojzis S.J. *et al* 1996. "Evidence for life on earth before 3,800 million years ago". *Nature* 384 (6604): 55–59.doi:10.1038/384055a0.
4. Stuart Wolpert 2015. Life on Earth likely started at least 4.1 billion years ago — much earlier than scientists had thought. UCLA newsroom. newsroom.ucla.edu. Accessed on 08-28-2016.
5. Elizabeth AB, Patrick B, Mark TH, Wendy LM. Potentially biogenic carbon preserved in a 4.1 billion-year-old zircon. PNAS Vol. 112 no. 47. pp 14518–14521.
6. Oparin, Alexander Ivanovich 2003. *The Origin of Life*. Courier Dover.
7. Knoll, Andrew H. 2004. *Life on a young planet: the first three billion years of evolution on Earth*. Princeton, N.J.
8. Martin, William; Russel, Michael J. (2003). "On the origins of cells: a hypothesis for the evolutionary transitions from abiotic geochemistry to chemoautotrophic prokaryotes, and from prokaryotes to nucleated cells". *Phil. Trans. R. Soc. B* **358**(1429): 59–85. doi:10.1098/rstb.2002.1183.
9. Scharf, Caleb; et al. (18 December 2015). "A Strategy for Origins of Life Research". *Astrobiology (journal)*. **15** (12): 1031–1042. doi:10.1089/ast.2015.1113. Retrieved 20 December 2015.
10. Peretó, Juli (2005). "Controversies on the origin of life" (PDF). *International Microbiology*. Barcelona: Spanish Society for Microbiology. **8** (1): 23–31.
11. Theobald DL (May 2010). "A formal test of the theory of universal common ancestry". *Nature*. **465** (7295): 219-22.Bibcode:2010Natur.465..219T. doi:10.1038/nature09014.
12. Robertson, Michael P.; Joyce, Gerald F. (May 2012). "The origins of the RNA world". *Cold Spring Harbor Perspectives in Biology*. Cold Spring Harbor, NY: Cold Spring Harbor Laboratory Press. **4** (5): a003608.doi:10.1101/cshperspect.a003608.

13. Cech, Thomas R. (July 2012). "The RNA Worlds in Context". *Cold Spring Harbor Perspectives in Biology*. Cold Spring Harbor, NY: Cold Spring Harbor Laboratory Press. **4** (7): a006742. doi:10.1101/cshperspect.a006742. ISSN 1943-0264.
14. Miller, Stanley L. (1953). "Production of Amino Acids Under Possible Primitive Earth Conditions" (PDF). *Science*. **117**(3046): 528–9. Bibcode:1953Sci...117..528M. doi:10.1126/science.117.3046.528.
15. Bada, Jeffrey L. (2000). "Stanley Miller's 70th Birthday" (PDF). *Origins of Life and Evolution of the Biosphere*. Netherlands: Kluwer Academic Publishers. **30**: 107–12. doi:10.1023/A:1006746205180.
16. Bada, Jeffrey L. (2013). "New insights into prebiotic chemistry from Stanley Miller's spark discharge experiments".*Chemical Society Reviews*. **42** (5): 2186–96. doi:10.1039/c3cs35433d.
17. Keller, Markus A.; Turchyn, Alexandra V.; Ralser, Markus (25 March 2014). "Non-enzymatic glycolysis and pentose phosphate pathway-like reactions in a plausible Archean ocean". *Molecular Systems Biology*. Heidelberg, Germany: EMBO Press on behalf of the European Molecular Biology Organization. **10** (725). doi:10.1002/msb.20145228.
18. Orgel, Leslie E. (April 2003). "Some consequences of the RNA world hypothesis". *Origins of Life and Evolution of the Biosphere*. Kluwer Academic Publishers. **33** (2): 211–218. doi:10.1023/A:1024616317965.
19. Cohen, B. A.; Swindle, T. D.; Kring, D. A. (2000). "Support for the Lunar Cataclysm Hypothesis from Lunar Meteorite Impact Melt Ages". *Science*. **290** (5497): 1754–1755. Bibcode: 2000Sci...290.1754C. doi:10.1126/science.290.5497.1754.PMID 11099411.
20. Sojo V, Herschy B, Whicher A, Camprubí E, Lane N. The Origin of Life in Alkaline Hydrothermal Vents.Astrobiology. 2016 Feb; 16 (2):181-97. doi: 10.1089/ast.2015.1406. Epub 2016 Feb 3.PMCID: PMC1693102.
21. William Martin and Michael J Russell. On the origins of cells: a hypothesis for the evolutionary transitions from abiotic geochemistry to chemoautotrophic prokaryotes, and from prokaryotes to nucleated cells.Philos Trans R SocLond B Biol Sci. 2003 Jan 29; 358(1429): 59–85.doi: 10.1098/rstb.2002.1183.
22. Russell, M.J., Barge, L.M., Bhartia, R., Bocanegra, D., Bracher, P.J., Branscomb, E., Kidd, R., McGlynn, S.E., Meier, D.H., Nitschke, W., Shibuya, T., Vance, S., White, L., &Kanik, I. (2014) The drive to life on wet and icy worlds. Astrobiology 14, 308-343.
23. Günter Wächtershäuser, G (1992). "Groundworks for an evolutionary biochemistry: The iron-sulphur world". *Progress in Biophysics and Molecular Biology*. 58 (2): 85–201. doi:10.1016/0079-6107(92)90022-X. PMID 1509092. Retrieved2009-05-02.
24. Ignatov I, Mosin OV (2013). Possible Processes for Origin of Life and Living Matter with Modeling of Physiological Processes of Bacterium Bacillus Subtilis in Heavy Water as Model System. Journal of Natural Sciences Research 3 (9), 65-76.

25. Ignatov I. (2010).Which water is optimal for the origin (generation) of life. Euromedica, Hanover, 34-37.
26. Ward, David M., et al. "A natural view of microbial biodiversity within hot spring cyanobacterial mat communities." *Microbiology and Molecular Biology Reviews* 62.4 (1998): 1353-1370.
27. Kurihara K, Okura Y, Matsuo M, Toyota T, Suzuki K, Sugawara T. A recursive vesicle-based model protocell with a primitive model cell cycle. *Nature Communications* Article number: 8352doi:10.1038/ncomms9352.
28. Roldan A et al. 2015. Bio-inspired CO_2 conversion by iron sulfide catalysts under sustainable conditions. Chem. Commun. 51, 7501-7504; doi: 10.1039/C5CC02078F.
29. Gold, T. (1992). "The Deep Hot Biosphere". Proceedings of the National Academy of Sciences. 89 (13): 6045–9.Bibcode:1992PNAS...89.6045G. doi:10.1073/pnas.89.13.6045.
30. Wickramasinghe, Chandra (2011). "Bacterial morphologies supporting cometary panspermia: a reappraisal". International Journal of Astrobiology. 10 (1): 25–30.
31. Hoyle, F. and Wickramasinghe, N.C., 1981. Evolution from Space (Simon & Schuster Inc., NY, 1981 and J.M. Dent and Son, Lond, 1981), ch3 pp. 35-49.
32. Lynn J. Rothschild; Adrian M. Lister (June 2003). *Evolution on Planet Earth – The Impact of the Physical Environment*. Academic Press.
33. Arrhenius, S., *Worlds in the Making: The Evolution of the Universe*. New York, Harper & Row, 1908.
34. Fred Hoyle; Chandra Wickramasinghe & John Watson (1986). *Viruses from Space and Related Matters*. University College Cardiff Press.
35. Wickramasinghe, D. T, Allen, D. A. (1983). "Three components of 3?4 ?m absorption bands". *Astrophysics and Space Science*. 97 (2): 369–378.
36. Weaver, Rheyanne (April 7, 2009). "Ruminations on other worlds".*statepress.com*. Retrieved 25 July 2013.
37. Than, Ker (25 April 2011). "Bacteria Grow Under 400,000 Times Earth's Gravity". *National Geographic - Daily News*. National Geographic Society. Retrieved 28 April 2011.
38. McLean, Robert JC, Allana K. Welsh, and Valerie A. Casasanto. "Microbial survival in space shuttle crash." *Icarus* 181.1 (2006): 323-325.
39. Parag A. Vaishampayan, ElkeRabbow, Gerda Horneck, and Kasthuri J. Venkateswaran (May 2012). "Survival of *Bacillus pumilus* Spores for a Prolonged Period of Time in Real Space Conditions". *Astrobiology*. 12 (5): 487–497.
40. Top Questions: Questions About Intelligent Design: What is the theory of intelligent design?". *Center for Science and Culture*. Seattle, WA: Discovery Institute. Retrieved 2012-06-16.
41. Hoyle, Fred, *Evolution from Space*, Omni Lecture, Royal Institution, London, 12 January 1982; *Evolution from Space* (1982) pp. 27–28.
42. Charles B. Thaxton, Walter L. Bradley, Roger Olsen 1984. The mystery of life's origin: Reassessing current theories.
43. Johnson, Phillip E. (2010). *Darwin on Trial, 3rd ed*. Downers Grove, Ill. InterVarsity Press.

44. Simon Conway Morris 2006. Darwin's dilemma: the realities of the Cambrian 'explosion'. Philos Trans R SocLond B Biol Sci. 2006 Jun 29; 361(1470): 1069–1083.Published online 2006 May 3. doi: 10.1098/rstb.2006.1846.
45. John van Wyhe, ed. 2002. *The Complete Work of Charles Darwin Online.* http://darwin-online.org.uk/.
46. Behe, Michael J. (1996). *Darwin's Black Box: The Biochemical Challenge to Evolution.* New York: Free Press.
47. Matzke, Nick (November 23, 2004). "Critique: 'Of Pandas and People'". *National Center for Science Education* (Blog). Berkeley, CA: National Center for Science Education.
48. Behe, Michael (1997). "Molecular Machines: Experimental Support for the Design Inference". Apologetics.org. Trinity, FL: The Apologetics Group; Trinity College of Florida. Retrieved 2016-04-18.
49. Gonzalez, Guillermo; Richards, Jay W. (2004). *The Privileged Planet: How Our Place in the Cosmos is Designed for Discovery.* Washington, D.C.: Regnery Publishing.
50. Dembski, William A. (1998). *The Design Inference: Eliminating Chance through Small Probabilities.* Cambridge; New York :Cambridge University Press.
51. Hoyle, Fred (1984). *The Intelligent Universe.* Holt, Rinehart, and Winston.
52. Ayala, Francisco J. (2007). *Darwin's Gift to Science and Religion.* Washington, D.C.: Joseph Henry Press.
53. Isaacson, Walter (2008). *Einstein: His Life and Universe.* New York: Simon and Schuster, pp. 390.

Chapter 6

Stellar life cycles: Is there life out there?

A distance of 4.24 light years separates the Sun from its closest stellar neighbor- Proxima Centauri.It's difficult to completely appreciate the enormity of such vast distances, but a popular analogy sets the Sun at the size of a grapefruit. If you wanted to get from your grapefruit sized Sun to the kumquat sized Proxima Centauri, you would have to travel approximately 2,500 miles, which is about the distance from coast to coast on the continental United States. Considering that this is just the distance to the Sun's closest neighbor within our galaxy, the vastness of the universe itself becomes extremely difficult to fully comprehend.

The observable universe consists of galaxies and all other forms of matter that can be observed from the Earth at present time. The detectable light and signals from the observable universe have traveled to reach Earth from the time of the cosmological expansion. Edwin Hubble provided evidence, through the study of the redshift of galaxies, that the universe is actually expanding. A redshift occurs whenever a light source moves away from an observer.This causes an increase in the wavelength of light or electromagnetic radiation emitted by an object, the wavelength thus shifts towards the red (longer wavelength)end of the electromagnetic spectrum. Since the universe is expanding the obvious conclusion, is that it began from one point, hence the Big Bang model of the universe. The phrase Big Bang was coined rather derisively by astronomer Fred Hoyle during a 1949 BBC Radio broadcast. He didn't subscribe to this theory of the origin of the universe hence his remark was an attempt to mock the theory.

The actual size of the universe is difficult to define, if we go by the theory of general relativity, there will exist regions of space that are unlikely to ever interact with other regions throughout the lifespan of the universe due to the ongoing expansion of the cosmos and the fixed speed of light. The spatial region of the cosmos that we can affect and vice versa is the observable universe. In reality, whether the universe beyond what is observable from our position in the Milky Way galaxy is finite or infinite is not known.

If we are to assume that the universe is uniform in all directions, then the distance to the edge of the observable universe is roughly the same in every direction. The observable universe is understood as a sphere around the Earth extending 93 billion light years across. This observable universe as described centers around the Earth, however the entire universe has no center. The observations made by cosmologists on the layout or arrangement of galaxies within the observable universe is intriguing. Galaxies are arranged in clusters, which form bigger groups called superclusters then walls and filaments. There also exist areas that are devoid of these structures called voids. Filaments are massive, thread-like formations with a length that typically ranges from 50 to 80 megaparsecs, the equivalent of 163 to 261 million light years, and form the boundaries between large voids in the universe. Other known notably sizable structures are the large quasar groups, which are in theory the precursors of galaxy filaments or superclusters. A quasar or quasi-stellar radio sources belong to a class of objects called **active galactic nuclei**. An active galactic nucleus is a compact region at the center of a galaxy that has a much higher than normal luminosity over at least some portion of the electromagnetic **spectrum. The luminosity of a quasar can be 100 times greater than that of the Milky Way. Before the Hubble space telescope, quasars were considered to be mysterious star-like objects that existed in isolation. With Hubble, several quasars observed have been found to all reside at the center of galaxies. Today most scientists believe that super massive black holes at the galactic centers are the "engines" that power the quasars.

Stars abound in our universe and form the building blocks of galaxies. Stars are responsible for the manufacture and distribution of heavy elements such as carbon, nitrogen, and oxygen, and their characteristics are intimately tied to the characteristics of the planetary systems that may coalesce about them. They have a life cycle that includes birth, a long lifespan and death.

Stars form within clouds of dust that are scattered across galaxies. Within these clouds, turbulence produces knots of gas and dust with sufficient gravitational pull to collapse on themselves. Following the collapse, the material at the center begins to heat up giving rise to a protostar, it is the hot core of the collapsing cloud that eventually becomes a star. As the cloud collapses, a dense, hot core forms and begins gathering dust and gas. Not all of this material ends up as part of a star- the remaining dust can become planets, asteroids, or comets or may remain as dust.

Three-dimensional computer models of star formation predict that the spinning clouds of collapsing gas and dust may break up into two or three blobs; this would explain why the majority the stars in the Milky Way are paired or exist in groups of multiple stars. Our closet neighbor, Proxima Centauri is considered to be part of a triple star system with Alpha Centauri A and B.

It takes approximately 50 million years for a star the size of our sun to mature from the start of the collapse of the clouds of gas and dust to adulthood. It will remain in the mature phase for about 10 billion years. Stars are powered by the nuclear fusion of hydrogen to form helium in their interiors. The outflow of energy from the central regions of the star provides the pressure necessary to keep the star from collapsing under its own weight, and the energy by which it shines. In general, the larger a star the shorter its life although, all but the most massive stars live for billions of years. When a star has fused all the hydrogen in its core, nuclear reactions come to a halt within its core. Without the energy production needed to support it, the interior begins to collapse into itself and becomes much hotter. Hydrogen however remains available outside the core, so hydrogen fusion continues in a shell surrounding this inner area. The increasingly hot center also pushes the outer layers of the star outward, causing them to expand and cool, transforming the star into a red giant. What happens next depends on the size of the core. Average stars, like our Sun will keep expelling its outer layers until its stellar core is exposed. This dead but intensely hot stellar body is called a white dwarf. White dwarfs are intrinsically very faint because they are so small and, lacking a source of energy production, they fade into oblivion as they gradually cool down.

The demise of stars may follow different pathways, white dwarfs may become novae if they occur in a binary or multiple star system.

If such a dwarf is close enough to a companion star, its gravity may drag matter- mostly hydrogen- from the outer layers of that star onto itself. By this it builds up its surface layer. When enough hydrogen has accumulated on the surface, a burst of nuclear fusion occurs, causing the white dwarf to brighten substantially and expel the remaining material. Within a few days, the glow subsides and the cycle starts again. Sometimes, particularly massive white dwarfs may accrete so much mass that they collapse and explode completely, becoming what is known as a supernova. Stars that are 8 times the mass of the sun or are even more massive are destined to end as supernovae.

Supernovae release an almost unimaginable amount of energy. For a period of days to weeks, a supernova may outshine an entire galaxy. On average, a supernova explosion occurs about once every hundred years in the typical galaxy. About 25 to 50 supernovae are discovered each year in other galaxies, but most are too far away to be seen without a telescope. If the collapsing stellar core at the center of a supernova contains between 1.4 and 3 solar masses, the collapse continues until electrons and protons combine to form neutrons, producing a neutron star. Neutron stars are incredibly dense. Because it contains so much mass packed into such a small volume, the gravitation at the surface of a neutron star is immense. Neutron stars also have powerful magnetic fields which can accelerate atomic particles around its magnetic poles producing powerful beams of radiation. Those beams sweep around like massive searchlight beams as the star rotates. If such a beam is oriented so that it periodically points toward the Earth, we observe it as regular pulses of radiation that occur whenever the magnetic pole sweeps past the line of sight. In this case, the neutron star is known as a pulsar.

If the collapsed stellar core is larger than three solar masses, it collapses completely to form a black hole: an infinitely dense object whose gravity is so strong that nothing can escape its immediate proximity, not even light.

The dust and debris left behind by novae and supernovae eventually blend with the surrounding interstellar gas and dust, enriching it with the heavy elements and chemical compounds produced during stellar death. Eventually, those materials are recycled, providing the building blocks for a new generation of stars and accompanying planetary systems.

At some point in our collective history, we believed rightly or wrongly that we had a unique place in the scheme of the universe. The evidence available then probably helped fuel this; 1) Although astronomers and philosophers believed that other stars might have planetary systems, there was unfortunately no evidence to show that planets and solar systems might be commonplace. 2) Mankind had through various space missions failed to discover any real evidence of life outside our planet. 3) The obvious fact that we had never had contact with any form of intelligent life. These all helped to support a notion that we might somehow be unique. But are we truly an accidental cosmological fluke?

Some scientists like Frank Drake believe the exact opposite to be true. Drake believed in not only the probability of just simple life being abundant but more importantly, intelligent life like ours. He is a notable pioneer in the search for extraterrestrial life. He developed an equation in 1961 which is called the Drake equation; which tries to arrive at an estimate of the number of active, communicative **extraterrestrial civilizations** in our galaxy. The number of such civilizations, N, is calculated by multiplying the following indices; 1) the average rate of star formation, R_*, in our galaxy, 2) the fraction of formed stars, f_p, that have planets, 3) the average number of planets per star that has planets, n_e, that can potentially support life, 4) the fraction of those planets, f_l, that actually develop life, 5) the fraction of planets bearing life on which intelligent, civilized life, f_i, has developed, 6) the fraction of these civilizations that have developed communications, f_c, i.e., technologies that release detectable signs into space, and 7) the length of time, L, over which such civilizations release detectable signals, for a combined expression of:

$$N = R_* \times f_p \times n_e \times f_l \times f_i \times f_c \times L$$

Even though the equation has received criticism for the fact that several of its terms are conjectural, with the net result being that the error associated with any derived value is very large such that the equation cannot be used to draw firm conclusions.

Whether firm conclusions can or cannot be drawn from the equation does not detract from the fact that the equation itself calls for robust thought on the plausibility of the existence of intelligent extraterrestrial life.

Recent discoveries strengthen the argument that there is a strong chance that intelligent life other than ours exists in the universe. For instance, the last three decades have witnessed the detection of numerous exoplanets, which are planets that orbit any star other than the Sun. Over 3,500 exoplanets have been discovered over this period of time, including 585 multiple planetary systems as of 1st October 2016. The Kepler space telescope has been at the forefront of discovering new exoplanets, having detected over two thousand of them since 2009. On average there is at least one planet per star, with a certain percentage having multiple planets. Sun-like stars with "Earth-sized" planets in the habitable zone have been found, and the discovery of these exoplanets has intensified interest in the search for extraterrestrial life. The special interest in planets that orbit a star's habitable zone stems from the possibility of liquid water existing on such planets. The presence of water would increase the chance of having life forms that are comparable to those on Earth. In November 2013, astronomers reported, based on Kepler space mission data, that there could be as many as 40 billion Earth-sized planets orbiting in the habitable zone of Sun-like stars and red dwarf stars in the Milky Way, 11 billion of which may be orbiting Sun-like stars. The exoplanet termed Kepler-186f for instance is similar in size to Earth with a 1.2 Earth-radius measure and it is located within the habitable zone around its red dwarf. This is just one of an increasing number of exoplanets with potentially ideal conditions for life.

Even before the discovery of exoplanets and planetary systems like ours, the search for extraterrestrial intelligence (SETI) had already begun. These new findings have served to provide added impetus to the ongoing search. **SETI is a broad based term for the various forms of the scientific search for intelligent life in locations other than our planet. The monitoring of radio signals for signs that they were** transmitted from civilizations on other worlds is an example. Focused international efforts to answer a variety of scientific questions concerning extraterrestrial intelligence have been going on since the 1980s. These efforts have yielded underwhelming results so far, however they did receive a significant boost in 2015, with Stephen Hawking and Russian billionaire Yuri Milner announcing a well-funded effort, called the Breakthrough Initiatives to expand efforts to search for extraterrestrial life. Breakthrough Initiatives is a program that is divided into multiple projects. **Breakthrough listen** will comprise an effort to search over 1,000,000 stars for artificial radio or laser signals. A parallel project called Breakthrough Message is an effort to create a message- representative of humanity and planet Earth. The project **Breakthrough Starshot** aims to send a swarm of probes to the nearest star at about 20% the speed of light.

Certain scientists like Enrico Fermi have expressed skepticism about the likelihood of intelligent life existing in abundance. This is highlighted in the Fermi Paradox, named after Enrico Fermi. The Fermi paradox is the apparent contradiction between the lack of evidence and high probability estimates for the existence of extraterrestrial civilizations. The paradox can be stated as follows; the size and age of the universe incline us to believe that many technologically advanced civilizations must exist. However, this belief seems logically inconsistent with our lack of observational evidence to support it. Either 1) the initial assumption is incorrect or technologically advanced intelligent life is much rarer than we believe, or 2) our current observations are incomplete and we simply have not detected them yet, or 3) our search methodologies are flawed and we are not searching for the correct indicators.

A variety of explanations proposed for the Fermi paradox range from analyses suggesting that intelligent life is rare- the rare Earth hypothesis- to analyses suggesting that although extraterrestrial civilizations may be common, they would not communicate, or would not travel across interstellar distances.

In **planetary astronomy** and **astrobiology**, the Rare Earth Hypothesis argues that the origin of life and the evolution of biological complexity such as sexually reproducing **multicellular organisms** on **Earth** and, subsequently **human intelligence** required an improbable combination of **astrophysical** and geological events and circumstances. The hypothesis argues that complex extraterrestrial life is a very improbable phenomenon and likely to be extremely rare. The geologist Peter Ward and astrobiologist Donald E. Brownlee are the originators of this hypothesis.

An alternative view is canvassed by scientists like Carl Sagan and Frank Drake. They argue that Earth is a typical rocky planet in a typical planetary system, located in a non-exceptional region of a common barred-spiral galaxy.

My opinion on the existence or non-existence of intelligent extraterrestrial life is based on the extremely ordered nature of the cosmos and the fact that most events in the universe follow a predictable pattern hence science was able to accurately predict the existence of black holes, gravitational waves, exoplanets and planetary systems, long before these were actually detected via improved technologies. "Are we alone?" Is a question that is relevant but absolutely unnecessary, just like the stars- that are birthed, mature and die- it is my personal conviction that we are most definitely not alone. Anything to the contrary, smacks of an obsession to shun open minded logic that recognizes the obvious- that if there are ubiquitous numbers of stellar "ecosystems" that are almost identical to our solar system- why then would intelligent life be an exclusive signature located on only one planet in the universe? That would be equivalent to viewing the universe as simply a massive wasteland, that somehow throws up conditions that simulate those in our solar system and planet but with a difficult to conceive proviso that life continually fails to take off in these other worlds. This is definitely not in line with any pattern of straightforward analytical reasoning. I will take a moment to point out that my opinion is by no means a rebuke or an attempt to ridicule those with a differing point of view. Not at all, as what would science be without varying hypotheses and opinions?

If humanity does not self destruct given the prospect of nuclear warfare, the technologies that will emerge within the next two centuries are likely to prove the existence of intelligent life outside our home planet and will make us look back and see how empty the question of possibly being alone in the universe was in the first place. To quote Stephen Hawking, "In an infinite universe, there must be other life. . ."

I agree 100%.

Is the Universe brimming with life or are we alone?

Reference

1. "Our local galactic neighborhood". NASA. February 8, 2000. Accessed on 02-04-2015.
2. Glister, Paul (September 1, 2010). "Into the interstellar void". Centauri Dreams. Retrieved March 22, 2013.
3. Gott III, J. Richard *et al* (2005). "A map of the Universe". *The Astrophysics Journal* **624** (2): 463. doi:10.1086/428890.
4. Davis, Tamara M, Charles H. Lineweaver (2004). "Expanding confusion: common misconceptions of cosmological horizons and the superluminal expansion of the universe". *Publications of the Astronomical Society of Australia* **21** (1): 97.doi:10.1071/AS03040.
5. Zeilik, Michael, Gregory, Stephen A. (1998). *Introductory Astronomy & Astrophysics* (4th ed.). Saunders College Publishing.
6. NASA/WMAP Science Team (24 January 2014). "Universe 101: Will the Universe expand forever?". NASA. Retrieved16 April 2015.
7. Brian Greene (2011). *The Hidden Reality*. Alfred A. Knopf.
8. Greenstein, Jesse L, Schmidt, Maarten (1964). "The Quasi-Stellar Radio Sources 3C 48 and 3C 273". *The Astrophysical Journal*. **140**: 1.
9. Shields, Gregory A. (1999). "A BRIEF HISTORY OF AGN". *The Publications of the Astronomical Society of the Pacific*. **111**(760): 661–678.
10. Richmond, Michael. "Late stages of evolution for low-mass stars". Rochester Institute of Technology. Accessed on April 8th2006.
11. Seligman, Courtney. "Slow Contraction of Protostellar Cloud". *Self-published*. Archived from the original on 2008-06-23. Accessed on Nov 4th2006.
12. Bally, J, Morse, J,Reipurth, B. (1996). "The Birth of Stars: Herbig-Haro Jets, Accretion and Proto-Planetary Disks". In Benvenuti, Piero; Macchetto, F. D.; Schreier, Ethan J. *Science with the Hubble Space Telescope – II. Proceedings of a workshop held in Paris, France, December 4–8, 1995*. Space Telescope Science Institute. p. 491.
13. Beech, M. *Alpha Centauri: Unveiling the Secrets of Our Nearest Stellar Neighbor*. New York: Springer, 2015. pp. x-xi.
14. Liebert, J. (1980). "White dwarf stars". *Annual Review of Astronomy and Astrophysics*. **18** (2): 363–398.
15. NASA. Stars. http://science.nasa.gov/astrophysics/focus-areas/how-do-stars-form-and-evolve/. Accessed on 08-03-2016.
16. Pranab Ghosh, *Rotation and accretion powered pulsars*. World Scientific, 2007,
17. Wald, Robert M. (1992). *Space, Time, and Gravity: The Theory of the Big Bang and Black Holes*. University of Chicago Press.
18. Stahler S. W, Palla F. (2004). *The Formation of Stars*. Weinheim: Wiley-VCH.
19. Burchell, M.J. (2006). "W(h)ither the Drake equation?". *International Journal of Astrobiology*. **5** (3): 243–250.

20. Schneider, Jean (30 August 2016). "Interactive Extra-solar Planets Catalog". *The Extrasolar Planets Encyclopedia*. Accessed 2016-08-30.
21. Lammer H, Bredehöft J. H, Coustenis A, Khodachenko M. L, et al. (2009). "What makes a planet habitable?" (PDF). *The Astronomy and Astrophysics Review*. **17**: 181–249.
22. Turnbull, Margaret C.; Tarter, Jill C. (March 2003). "Target selection for SETI: A catalog of nearby habitable stellar systems" (PDF). *The Astrophysical Journal Supplement Series*. **145**: 181–198.
23. Petigura Erik A, et al. (31 October 2013). "Prevalence of Earth-size planets orbiting Sun-like stars". *Proceedings of the National Academy of Sciences of the United States of America*.
24. SETI at 50". *Nature*. **416** (7262): 316. 2009.
25. Lendino, Jamie (20 July 2015). "Stephen Hawking, Milner unveil $100M initiative to 'dramatically accelerate' search for alien life". ExtremeTech. Accessed on July 13th 2015.
26. Gilster, Paul (12 April 2016). "Breakthrough Starshot: Mission to Alpha Centauri". *Centauri Dreams*. Retrieved 14 April 2016.
27. Chris Impe (2011). *The Living Cosmos: Our Search for Life in the Universe*. Cambridge University Press.
28. Lineweaver, Charles H (2008). *Paleontological tests: human-like intelligence is not a convergent feature of evolution*. From fossils to astrobiology. Springer. pp. 353–368.
29. Ward, Peter D.; Brownlee, Donald (January 14, 2000). *Rare Earth: Why Complex Life is Uncommon in the Universe (1st ed.)*. Springer.
30. Webb, Stephen (2002). *Where is Everybody? (If the universe is teeming with aliens, Where is Everybody?: Fifty solutions to the Fermi paradox and the problem of extraterrestrial life)*. Copernicus Books (Springer Verlag).
31. Sagan, Carl (1985) [Originally published 1980]. *Cosmos* (1st Ballantine Books ed.). New York: Ballantine Books.

Chapter Seven

The end of time: what is the future of the cosmos?

To have any idea concerning the future of the universe, it is essential to look to its beginnings.

When did time begin or does time have a beginning? The Big Bang model of the universe has it that time actually does have a beginning. The Big Bang Theory is the leading explanation about how the universe began. It tries to explain the start of the universe by using extrapolation of the expansion of the universe going backward in time using Einstein's theory of general relativity. The universe is known to be expanding, which means if we were to put time in reverse the universe would start to shrink, like rewinding the video of a balloon being inflated. This backward analysis, reaches a point with an infinite density and temperature at a finite time in the past. This would mean a singularity with the breakdown of all the laws of physics including general relativity. A singularity is a point in which all physical laws are indistinguishable from one another, where space and time are no longer interrelated realities. This singularity is sometimes called "the

Big Bang" but the term can also refer to the early hot, dense phase itself, which can be considered to be the "birth" of our universe.

Based on measurements of the expansion using type Ia supernovae and measurements of temperature fluctuations in the cosmic background radiation, the universe has an estimated age of 13.8 billion years. The detailed structure of the cosmic microwave background fluctuations depends on the current density of the universe, the composition of the universe and its expansion rate. As of 2013, the Wilkinson Microwave Anisotropy Probe (WMAP) determined these parameters with great accuracy. The key here is that by knowing the composition of matter and energy density in the universe, we can use Einstein's General Relativity to compute how fast the universe has been expanding in the past. With this information, we can turn the clock back and determine when the universe had "zero" size. The time between then and now is the age of the universe. The agreement of these independent measurements strongly supports the lambda cold dark matter (ΛCDM) model of the Big Bang that describes in detail the contents of the universe.

The cosmic background radiation commonly called the cosmic microwave background radiation (CMBR) is radiation left over from the early development of the universe, the afterglow of the Big Bang.

Going by the theories of physics, if we were to take a look at the universe one second after the Big Bang, what we would see is a 10 billion degree sea of neutrons, protons, electrons, positrons, photons, and neutrinos. Then, as time goes on we would observe the universe cool, the neutrons either decaying into protons and electrons or combining with protons to make nuclei of deuterium- an isotope of hydrogen. As it continued to cool, it would eventually reach the temperature where electrons combined with nuclei to form neutral atoms. Before this "recombination" occurred, the universe would have been opaque because the free electrons would have caused light (photons) to scatter the way sunlight scatters from the water droplets in clouds. But when the free electrons were absorbed to form neutral atoms, the universe suddenly became transparent. Those same photons constitute the afterglow of the Big Bang known as the CBMR.

The Big Bang is not an explosion of matter moving outward to fill an empty universe. Instead, **space itself expands** with time everywhere and increases the physical distance between two comoving points. In other words, the Big Bang is not an explosion in space, but rather an expansion of space.

Over a long period of time, the slightly denser regions of the nearly uniformly distributed matter, gravitationally attracted matter that was nearby and thus grew even denser. Eventually gas clouds, stars, galaxies, and the other astronomical structures observable today formed. The details of this process depend on the amount and type of matter in the universe. The four possible types of matter are known as **cold dark matter, warm dark matter, hot dark matter,** and **baryonic matter**. Baryonic matter includes only matter composed of baryons. A baryon is a composite subatomic particle made up of three quarks. In other words, it should include protons, neutrons and all the objects composed of them, but exclude things such as electrons and neutrinos which are actually leptons. Dark matter on the other hand is considered to be the non-luminous material which exists in space and which could take either of two forms: weakly interacting particles; *cold dark matter*, or high-energy randomly moving particles created soon after the Big

Bang; *hot dark matter*. It can be detected through its gravitational effects.

The best measurements available from **WMAP** show that the data is well-fit by a Lambda-CDM model in which dark matter is assumed to be cold, and is estimated to make up about 23% of the matter/energy of the universe, while baryonic matter makes up about 4.6%.

Independent lines of evidence from type Ia supernovae and the **CMBR** imply that the universe today is dominated by a mysterious form of energy known as **dark energy**, which apparently permeates all of space. **Supernovae observations showed that the expansion of the Universe, rather than slowing, is accelerating. Something, not like matter and not like ordinary energy, is pushing the galaxies apart.** The observations suggest 73% of the total energy density of today's universe is in this form. When the universe was very young, it was likely infused with dark energy, but with less space and everything closer together, gravity predominated, and was slowly braking the expansion. But eventually, following billions of years of expansion, the growing abundance/density of dark energy caused the **expansion of our universe** to slowly begin to accelerate. Dark energy in its simplest formulation takes the form of the cosmological constant term in Einstein's field equations of general relativity, but its composition and mechanism are unknown and, it continues to be investigated both theoretically and through observation.

The cosmic events following the inflationary epoch can be described and modeled by the ΛCDM model of cosmology. This model does so by combining aspects of quantum mechanics and general relativity. The inflationary epoch was the period in the evolution of the early **universe** when the universe underwent an extremely rapid exponential expansion. The expansion is thought to have been triggered by the **phase transition** that marked the end of the preceding *grand unification epoch* at approximately 10^{-36} seconds after the Big Bang. One of the theoretical products of this phase transition was a **scalar field** called the **inflation** field. As this field settled into its lowest energy state throughout the universe, it generated a repulsive force that led to a rapid expansion of space. This expansion explains various properties of the current universe that are difficult to account for without such an inflationary epoch.

However prior to 10^{-15} seconds, the events are not sufficiently described nor explained by any existing model, this may possibly be resolved by a unified theory of quantum gravitation. No such theory exists at this point in time. An understanding of the earliest period in the history of our galaxy is one of greatest unresolved problems in physics.

Will the universe go on to expand forever, will it stop or will it go into reverse and come back together in a massive crunch? The ultimate fate of the universe will depend upon the physical properties of the mass/energy in the universe, its average density, and the rate of expansion.

The density parameter, Omega (Ω) is an important parameter in fate of the universe; it is defined as the average matter density of the universe divided by a critical value of that density. This makes the universe take up one of three possible geometries depending on whether Ω is equal to, less than, or greater than 1. These are respectively called, the flat, open and closed universes. These three adjectives refer to the overall geometry of the universe, and not to the local curving of space-time caused by smaller clumps of mass for example galaxies and stars.

The closed universe model.
If Ω > 1, then the geometry of space is closed like the surface of a sphere, the sum of the angles of a triangle exceeds 180 degrees and there are no parallel lines; all lines eventually meet. The geometry of the universe is, at least on a very large scale, elliptic.
In a closed universe, gravity eventually stops the expansion of the universe, after which it starts to contract until all matter in the universe collapses to a point; a final singularity termed the "Big Crunch", which is technically the opposite of the Big Bang. If, however, the universe contains a significant amount of dark energy as suggested by recent findings, its repulsive force may be sufficient to cause the expansion of the universe to continue forever, even if Ω > 1.

The open universe model.
If Ω < 1, the geometry of space is open, i.e., negatively curved like the surface of a saddle. The geometry of such a universe is **hyperbolic**. Even without dark energy, a negatively curved universe expands forever, with gravity negligibly slowing the rate of expansion. With dark energy, the expansion not only continues but accelerates. The ultimate fate of an open universe is either universal heat death- the "Big Freeze", or the "Big Rip", where the acceleration caused by dark energy eventually becomes so strong that it completely overwhelms the effects of the gravitational, electromagnetic and weak nuclear binding forces.

The flat universe model.
If the average density of the universe exactly equals the critical density so that Ω = 1, then the geometry of the universe is flat. Measurements from the **Wilkinson Microwave Anisotropy Probe** have confirmed the

universe is flat with only a 0.4% margin of error. In this model the universe would have the same fate as the open universe.

It is obvious that further advances in physics are essential before the ultimate fate of the universe can be predicted with any level of certainty. At the moment the three scenarios described are possible end points. The most plausible is that the universe is flat.

The Big Crunch, which can be looked upon as the reverse of the Big Bang, a collapsing universe would entail an initial decrease in size that occurs evenly as matter is distributed in a reasonably consistent manner. At the start, the rate of contraction would be slow, but the rate would gradually increase. The temperature begins to increase exponentially, stars would explode and vaporize. On a microscopic scale, atoms and even nuclei would disintegrate in a reverse mirroring of the early stages of the Big Bang. The collapse will continue until a singularity is reached. According to some predictions, very close to the singularity the warpage of space-time will be extremely violent and chaotic. This model offers the possibility of a cyclic universe or Big Bounce. The events of the crunch would be succeeded by rebirth of the universe through the Big Bang, this could continue forever. However the recent evidence of an accelerating universe makes this a less than likely outcome.

The Big Rip on the other hand is a situation that may theoretically occur in an ever expanding universe. In this model, at a certain point during the expansion of the cosmos, galaxies would be separated from each other. Approximately sixty million years before the end, gravity would be too weak to hold the individual galaxies, like the Milky Way, together. About three months before the end, planetary systems would be gravitationally unbound. In the final minutes, stars and planets would be torn apart. Atoms would be destroyed an instant before the end, with the fabric of the universe itself being ripped apart at the very end.

The most likely end point, based on current knowledge, is that of a long and slow decline, the Big Freeze or Heat Death. In this scenario, the universe continues to expand and gradually attains a state of zero thermodynamic free energy, in which it is unable to sustain motion or life. Eventually, over a time scale of a hundred trillion years or more, it would reach a state of maximum entropy at a temperature of very close to absolute zero, where the universe simply becomes too cold, and all that would remain are burned-out stars, cold dead planets and black holes.

What happens after that is even more speculative but, eventually, even the atoms making up the remaining matter would start to degrade and disintegrate, as protons and neutrons decay into positrons and electrons, which over time would collide and eradicate each other. Depending on the rate of expansion of the universe at that time, it is possible that some electrons and positrons may form bizarre atoms known as positronium. After perhaps over multiple trillions of years, even the positronium will have collapsed and the constituent particles will annihilate each other.

In this way, all matter would slowly evaporate away as energy, leaving behind only black holes, ever more widely dispersed as the universe continues to expand. The black holes themselves would break down eventually, slowly leaking away *Hawking radiation*, until, after several more trillions of years, the universe will exist as just empty space and weak radiation at a temperature infinitesimally above absolute zero.

Hawking Radiation refers to electromagnetic radiation which, according to theory, should be emitted by a black hole. The radiation is due to the black hole capturing one of a particle-antiparticle pair created spontaneously near to the event horizon.

At the end of the universe, time itself will lose all meaning as there will be no events of any kind, and therefore no frame of reference to indicate the passage of time or even its direction.

The final fate of the Earth appears more predictable than that of the greater cosmos. The planet will ultimately vanish or be totally scorched within the next six to seven billion years when the Sun's atmosphere balloons outward to engulf it during its red giant phase.

How will it all end; a Big Crunch, a Big Freeze or a Big Rip...

Reference

1. Joseph Silk (2009). *Horizons of Cosmology*. Templeton Press.

2. Simon Singh (2005). *Big Bang: The Origin of the Universe*. Harper Perennial.

3. Chow, Tai L. (2008). *Gravity, Black Holes, and the Very Early Universe: An Introduction to General Relativity and Cosmology*. Springer.

4. Clavin, Whitney (17 March 2014). "NASA Technology Views Birth of the Universe". NASA. Retrieved 17 March 2014.

5. Frieman, Joshua A.; Turner, Michael S.; Huterer, Dragan (September 2008). "Dark Energy and the Accelerating Universe".*Annual Review of Astronomy and Astrophysics*. **46** (1): 385–432.

6. Hillebrandt, W.; Niemeyer, J. C. (2000). "Type IA Supernova Explosion Models". *Annual Review of Astronomy and Astrophysics*. **38** (1): 191–230.

7. "First minutes of the Big Bang". What is USA News. 12 March 2014. Retrieved 2013-11-19.

8. NASA. Our Universe; matter/energy.http://map.gsfc.nasa.gov/universe/uni_matter.html

9. Peebles, P. J. E. &Ratra, Bharat (2003). "The cosmological constant and dark energy". *Reviews of Modern Physics*. **75**(2): 559–606.

10. Frieman, Joshua A.; Turner, Michael S.; Huterer, Dragan (2008-01-01). "Dark Energy and the Accelerating Universe".*Annual Review of Astronomy and Astrophysics*. **46** (1): 385–432.

11. Durrer, R. (2011). "What do we really know about Dark Energy?". *Philosophical Transactions of the Royal Society A: Mathematical, Physical and Engineering Sciences*. **369** (1957): 5102.

12. Y Wang, J M Kratochvil, A Linde, and M Shmakova, *Current Observational Constraints on Cosmic Doomsday*. JCAP 0412 (2004).

13. Aurich, Ralf; Lustig, S.; Steiner, F.; Then, H. (2004). "Hyperbolic Universes with a Horned Topology and the CMB Anisotropy". *Classical and Quantum Gravity*. **21** (21): 4901–4926.

14. Ryden, Barbara. *Introduction to Cosmology*. The Ohio State University. p. 56.

15. WMAP - Fate of the Universe, *WMAP's Universe*, NASA. Accessed online July 17, 2008.

16. Tegmark, Max (2014). *Our Mathematical Universe: My Quest for the Ultimate Nature of Reality* (1 ed.). Knopf.

17. Overduin, James; Hans-Joachim Blome; Josef Hoell (June 2007). "Wolfgang Priester: from the big bounce to the Λ-dominated universe". *Naturwissenschaften*. **94** (6): 417–429.

18. Ellis, George F. R., R. Maartens, and M. A. H. MacCallum. Relativistic Cosmology. Cambridge: Cambridge UP, 2012. 146-47.

19. Possible Ultimate Fate of the Universe, Jamal N. Islam, *Quarterly Journal of the Royal Astronomical Society* **18** (March 1977), pp. 3–8.

20. Adams, Fred C.; Laughlin, Gregory (1997). "A dying universe: the long-term fate and evolution of astrophysical objects". *Reviews of Modern Physics*. **69**: 337–372.

21. Krauss, Lawrence M.; Starkman, Glenn D. (2000). "Life, the Universe, and Nothing: Life and Death in an Ever-expanding Universe". *Astrophysical Journal*. **531**: 22–30.

22. Schröder, K.-P.; Connon Smith, Robert (2008), "Distant future of the Sun and Earth revisited", *Monthly Notices of the Royal Astronomical Society*, **386** (1): 155–163.

Conclusion

Footprints in the sands of time.

There are gaps in our current body of scientific knowledge, such as in the laws governing the universe or the question of how life began. These two are just a few of a plethora of unanswered or unresolved questions.

It is encouraging that there has been an explosion of knowledge since the early twentieth century, with the pace of discovery of new information and scientific findings propelling our race forward with every passing day.

The future appears to have possibilities that are endless; the use of robotics to overcome paraplegia, interactive houses equipped with artificial intelligence so that it communicates and regulates other intelligent devices that communicate with one other as well as with humans, manned space missions to Mars and far beyond, scientists being able to switch off the process of aging, teleportation, the use of nanotechnology to keep humans in perpetual good health, and these are just a few of the wonders that are to become our reality in the future. A few of these innovations are already available.

As our civilization advances will we become more or less religious? Religion in my opinion will always be around in some form no matter how technologically advanced we get. Dismiss sentiments for a moment and observe how despite the tremendous strides we have made in all the sciences, religion has continued to flourish. This is why my opinion is strongly tipped in favor of the notion that we are indeed wired to search for a Creator.

A contrary argument might be the dwindling numbers of people attending orthodox churches in a country like England for example. Such an argument fails to factor in the rise of Pentecostal churches in England with their surging numbers in membership. If the numbers on either side of this divide are tallied, a more accurate picture begins to emerge. Christianity is not on a decline.

Current scientific knowledge gives us approximately a four percent understanding of our universe. So there is a whole lot that we all are completely ignorant about.

Does a Creator exist? I say an unequivocal yes, this I say with a sense of humility knowing that there will always be phenomena that will never be explained by any equation, theory or observation.

We all know there are laws that govern the universe, science has been able to identify some of them, I consider it an act of reckless warped logic to assume these laws developed randomly or on their own. This would be equivalent to saying that the metropolitan areas of our planet auto-created their own skyscrapers and electricity. Not a plausible argument at all.

Using medical science to critique this form of logic, let us use the following scenario; as a doctor when do you actually say a patient has died? Is there a difference between clinical death and actual death? Are we dead when we stop breathing- no respiratory movements are observable, our hearts stop beating- asystole on an EKG, our brains cease to have activity, our pupils are fixed and non-responsive to light? Could be, should be, but unfortunately this is not always the case. There have been so many cases of those who had been certified dead coming back to life with no resuscitation given. Were they dead or comatose? Could there be a phrase such as a *"temporarily suspended state of life defining activity,"* I think it is a strong possibility that this is different from clinical death that responds to cardiopulmonary resuscitation. What is the true dividing line between life and death? When should life support be switched off for a patient in a coma? A week, a month or may be a year? A Polish railway worker, Jan Grzebski was in a coma for 19 years but eventually regained consciousness. During this period he was mostly nursed by his loving wife with the antiquated (by today's benchmark) healthcare facilities that existed in his homeland at the time. Why and how was he able to survive, regain consciousness and go on to lead a normal life? The world may never truly know, we can study him in a lab for eternity and still fail to get the right answers.

Brain imaging techniques are wonderful but still do not fully answer all the questions. There are individuals with brain damage that is so extensive that their prognosis is either certain death or a vegetative state, yet they somehow reawaken and go on to live a normal life. In medical science too, just like other sciences there are still a lot of grew areas.

The conflict between scientific reasoning and religious faith based analysis will continue until the scientific community comes to a convincing conclusion as to the existence or non-existence of a Creator. The paradox here remains that both science and religion seek to provide answers to similar questions. Their methods may vary greatly but they have a wide spectrum of common interests- providing explanations and helping us truly understand our purpose.

The truth is that, knowledge breeds a certain amount of individual as well as collective arrogance that fosters a non-tolerance for dissenting opinion. It takes a combination of great insight and an open mind to accommodate and analyze points of view that are at variance with what we consider to be sacrosanct.

What is the ultimate fate of humanity, will we become extinct from our own parochial tendency towards intolerance; be it religious, ideological, economic or technological? Will we develop artificial intelligence that will turn against us? Will humanity adapt, thrive and flourish? Like science, even the future has so many unresolved uncertainties.

I believe our race has shown amazing resilience and will find a way to navigate new and emerging threats to its existence. I believe manned interplanetary travel closely followed by interstellar expeditions are the next frontiers, intergalactic space exploration lies much farther down in our common future as our aim remains to firmly plant our footprints in the sands of cosmic time.

www.ingramcontent.com/pod-product-compliance
Lightning Source LLC
Chambersburg PA
CBHW071820200526
45169CB00018B/502